MAKE IT SIMPLER

A Practical Guide to
Problem Solving in Mathematics

Carol Meyer and Tom Sallee

DALE SEYMOUR PUBLICATIONS

Pearson Learning Group

Design: Betsy Bruneau Jones
Illustrations: Elizabeth Callen

ISBN 0-201-20036-8
Printed in the United States of America
25 26 27 28 29   08 07 06 05 04

Dale
Seymour
Publications
Pearson Learning Group

1-800-321-3106
www.pearsonlearning.com

To our parents —
Ralph and Myrtle Johnson
George and Mary Sallee —
who showed us how to solve problems,
and to all of our students,
who showed us how to teach others to do so.

Acknowledgements: This book owes much to the many friends all over the country who have tried out different parts or whole versions of the text and have shared their comments with us. We are particularly grateful to Alice Alford, Mark Cary, Jon Geil, Doris Hirschorn, and Wanda Lincoln for all of their assistance and insights.

CONTENTS

Preface

The increased presence of the computer and the overwhelming demand for programmers in the last decade brought to light a critical deficiency in the teaching of mathematics—many good students could not solve problems. Although they could add fractions, solve quadratic equations and even manipulate Venn diagrams with ease, they were helpless when confronted with new word problems.

In 1980, the National Council of Teachers of Mathematics crystalized the growing dissatisfaction with the current state of affairs and produced an excellent yearbook of problem solving to deal with it. Since then, problem solving has become the teaching touchstone of the 1980s—and well it should, as almost everyone agrees. But there the agreement ends and the real difficulty begins: HOW can you teach children to solve problems effectively?

We think that we know how. We have developed an approach over the past five years which has been used by friends from New York to California. It is easy to use, adaptable to almost any curriculum, and IT WORKS.

This book tells you *how* it works, *why* we think it works, and *what* makes it work: lots of problems, games and strategies. The design is both specific and adaptable.

The following features of the program are those which we consider particularly important.

1. We begin at the beginning—with motivating the students.

2. The program can be used in conjunction with your current mathematics curriculum in about fifteen minutes a day.

3. There is a carefully-coordinated, *year-long* development of the skills needed by successful problem-solvers.

4. Each problem comes in three levels of difficulty: easy, medium and hard. The problems look very different to the students, but each of the three has the same mathematical structure.

5. The program is *easy* to implement.

6. Students *learn* how to solve problems.

7. Students *enjoy* the activities.

8. Problem solving skills carry over into other areas.

9. When students repeat the curriculum in another grade, they still enjoy the program and learn even more.

Introduction

The approach you will find in this book is based on the following beliefs:

1. The developmental theorists (Piaget, Bruner, et al) are essentially correct in their understanding of how minds mature as children grow.

2. It follows from these theorists that *the ages ten to fourteen* are crucial in the development of adult thinking styles. If people learn proper methods of thinking during these years, these good habits will be an aid to them for the rest of their lives. On the other hand, if they learn improper thinking methods at this age, their problem-solving capabilities will be permanently impaired.

3. Children can be taught to reason well. Most of the research in this area has focused on one-to-one settings, but the fact itself has been established. Thus, what is needed is an *effective* way to teach a whole classroom full of children to reason well.

4. Learning to reason requires *interaction* with someone else. Very few individuals are able to critique their own reasoning, and this is especially true of children. Children, therefore, need as much opportunity as possible to test the correctness of their reasoning with others.

5. Learning a *relatively few* reasoning techniques will enable an individual to deal effectively with most types of problems. These techniques are valuable not only in mathematics, but in science, in social science and in English as well.

6. The major barrier to effective problem solving is *psychological* rather than intellectual. Many students are so accustomed to having someone else solve their problems, that they simply do not try to solve them on their own.

From these beliefs, we developed a two-pronged approach for children of the appropriate ages. First, almost all activities must take place in a group setting.

Second, the students are taught a sequence of necessary skills: understanding a problem, learning general strategies and selecting an appropriate strategy.

The group setting has many advantages for both the classroom teacher and the student. From the teacher's perspective, the number of students in the class is reduced from thirty-two to eight. Moreover, since members of the group respond to routine inquiries, more of the teacher's time can be spent dealing with questions of substance. The students also benefit because they have the continuous presence of someone to whom they can explain their ideas. They also get the opportunity to teach and to refine their ideas, questions and approaches in the security of a small group. Thus they become more apt to dare, because there is less to fear in failure.

Within these groups, the students learn the basics of solving problems, beginning with the idea that problem solving is not such a difficult skill after all. A wide variety of activities are used beginning with open questions and gradually moving toward questions which have a definite solution. Many of these activities are in the form of games to increase student interraction and hypothesis testing within the group.

When the students are comfortable within their groups and have acquired confidence and some skill in articulating their ideas, they begin to learn more formal problem-solving skills. Clearly, the first skill they need is how to understand a problem; to understand it, they must be able to read it. Students acquire the art of reading a problem by writing several problems of their own. As this work progresses, the students begin writing more and more complex stories, testing each other to uncover the essence of each problem. At the same time, through another type of game, students learn the art of organizing information in useful ways.

By this time, the students are ready to appreciate the idea of using general strategies to solve problems. Even if they are not aware of the fact, they have been successfully solving problems for four months and are ready to more on to more difficult, mind-expanding challenges. To cope with these problems successfully, they need a new set of tools—a few useful strategies to simplify problems, to change these problems in such a way to help find a solution. In this book we utilize five such strategies: finding a subproblem, drawing a picture, using smaller numbers, looking for patterns and working backwards. Each of them should be examined carefully, so that students learn the strengths of each approach.

It is *crucial* to their success during the year that the students examine only a few problems, but that they examine these thoroughly. Bitter experience has taught us that several problems quickly covered teach the average student less than one or two problems completely examined. Students need to see *why* one particular approach works better than another. They need time to consider different ways of looking at a problem, so that they can truly understand its underlying structure. The answer is really unimportant here. The quality of thinking is all. And this quality of thinking is improved by the discussions within the groups, by presentations to the whole class and by the modeling of good thinking by the stronger class members.

Once students have mastered the mechanics of problem solving—in particular, have learned the major techniques which may be used to simplify problems—then they need more practice in matching appropriate strategies to particular problems. To teach this skill, we use practice coupled, as always, with discussion within the groups and with the entire class. As students do more problems (but still, only a few per week) they will develop greater facility in selecting appropriate strategies.

If the program is truly successful, by the end of the year the students will have integrated these strategies into their thinking styles so thoroughly that few students will recognize when they are using a particular one. From being consciously selected strategies, the methods will become as natural, and as useful, as walking. Moreover, the students will be *good* problem solvers.

The preceding discussion has been to show the philosophical overview of the year. In a nutshell, students should first learn not to be afraid of problems, how to read them carefully, how to make problems simpler, and then they must practice selecting appropriate strategies. The back of this book contains 107 problems of all types and all levels of difficulty which may be reproduced for your class. (The author will gladly pay $5 and give credit to the author of any problem used in subsequent editions of the book.)

We will now move on to a more detailed outline of the topics and the timeline governing their introduction over the course of the school year.

Timeline for the Year

This is a suggested timeline that has been used successfully in classrooms. Each of the activities, along with the philosophy behind them, are discussed in subsequent chapters. It is hoped that after carefully reading the entire book before beginning this program, the teacher will be able to use this timeline as a quick review and reference for the year of activities ahead.

SEPTEMBER

First Week Introduction

Sitting in groups of four

Rules for groups of four

Review unit of study needing a partner

Suggestions:

A. These activities require using the entire math period *each day.*
B. Allow time for discussion of rules *each day.*
C. Practice cooperation in groups of four.

**SEPTEMBER/
OCTOBER**

Simultaneous Activities

Practice with groups of four

Classroom problem-solving activities

Curriculum problem-solving activities

Logic activities

Suggestions:

A. These activities may take the entire math periods some days, other days, only five or ten minutes.
B. Maintain a supportive environment.
C. Do one of the above activities everyday.
D. Do a minimum of two problem-solving activities each week.
E. Allow time for class discussions about feelings, procedures, etc.

NOVEMBER   *Basic Headlines & Stories*

Introduction of headline/story concept

Group-of-four stories and headlines

Individual stories and headlines

Group of four as editing group

Restate the story

Clarify the question

Check the solution with the problem

Originate class Problem-Solving Chart

"Make It Simpler"

Suggestions:

A. Some of these activities may take more than fifteen minutes; others will require less time.

B. Do at least one of these activities each day.

C. Occasionally review activities done earlier in the year, especially review of rules for groups of four and some logic activities.

DECEMBER   *Silent Board Games*

Single-digit patterns

Organizing data

Two-digit patterns

Organizing data

Check solution with problem

Maintain and add to class Problem-Solving Chart

Suggestions:

A. These may take less than fifteen minutes some days.

B. Do at least one *every* day.

C. Do two or three problem-solving activities during this month.

JANUARY   *Advanced Headlines & Stories*

Organize data

Find unneeded information

Find assumed information

Find needed information

Find the hidden question

Suggestions:

A. These activities should take about fifteen minutes a day.
B. Work on these at least four out of five days.
C. Randomly review logic activities, silent board games and group-of-four discussions.

FEBRUARY/
MARCH/APRIL

Solving The Problem

Find a subproblem

Draw a picture or use manipulatives

Use smaller, easier numbers, fewer steps

Find a pattern

Work backwards

Suggestions:

A. These activities require daily sessions of fifteen minutes each.
B. Spend two weeks on each strategy.
C. Refer back to previous strategies.
D. Maintain and continue class Problem Solving Chart.
E. Discuss decision-making procedures.
F. Utilize steps in Understanding the Problem.

APRIL/MAY

Decision-Making

Review all strategies

Introduce alternate strategies

Suggestions:

A. These activities require three to five sessions a week, fifteen to twenty minutes each time.
B. Conduct teacher-led discussions frequently.

Answers

112	Sharon pays Janet $2.
113	Bryan gets $4.50; Bill $1.50.
114	80 nuts
115	Both a Smith and a Jones child eat one pound per week.
123	Goose over; corn or dog over; goose back; dog or corn over; goose over.
124	Many solutions. One is PNDQ QDNP NPQD DQPN
125	Three blocks west, one north
126	Use two 7-inch and three 5-inch sticks.
132	5352 plates
133	6114 people
135	10 pennies, 1 nickel, 1 dime
143	8192 mice
144	453 feet
152	$12,000
153	Six people
161	Six 4¢ stamps and one 7¢ stamp or four 4¢ stamps and one 15¢ stamp
162	Four solutions; see page 102.
163	Three solutions; see page 104.
211	5 homes
212	Randy owes Amy $1; Sarah owes Amy $1; Sarah owes Dave $3; or Randy owes Dave $1; Sarah owes Amy $2; Sarah owes Dave $2.
213	60¢
214	20 tickets
215	About 330 paprika dogs per outlet vs 160 southern fried. Paprika better
221	8 lines
222	90 seconds
223	Two geese over; two corn over; two geese back; dog and one corn over; two geese over.
224	Many solutions. One is: P R T M A R A M P T M T R A P A M P T R T P A R M A = apple, P = peach, R = pear T = apricot, M = plum

225	15 inches on a side
226	Two 8 1/2 sides by an 11″ side
231	7613 gms
232	3552 gizzards
233	Part will be 17 cartons high.
234	3200 cubic feet
235	Three solutions: (40 pennies, 5 dimes; 40 pennies; 3 nickels, 1 quarter; 35 pennies, 9 nickels, 1 dime).
241	Sample answer: 25 students give 60° handshakes
242	1431 tubes
243	2,047,000
244	$2.05; $15.85
251	44 apples
252	$34
253	12 cups
261	16 solutions; see page 152.
262	21 combinations; there can be 0, 2, 4, 6, 8, 10, 12, 14 or 16 boats with 7 people.
263	Fill the 8 kg tin 10 times and the 5 kg tin twice; or the 8 kg tin 5 times and the 5 kg tin 10 times; or the 5 kg tin 18 times. If you allow pouring from one can to another, there are many, many solutions.
312	Randy gives $1 to Amy; Sarah gives $1 to Amy and $3 to Dave or Randy gives $1 to Dave; Sarah gives $2 to Amy and $2 to Dave.
313	If Brad is paid by the paper, he should get 40¢.
314	24 mints
315	Each boy uses $\frac{13}{120}$ bottle more per week
322	45 seconds
323	All Johnsons and Rivers over; Rivers back; two Smiths and two Harrisons over, all Johnsons back; two other Harrisons and Rivers over; Rivers back; all Johnsons and Rivers over
324	See page 170.
325	See page 172.
326	See page 174.
332	137,195 cubic meters
333	65.7 minutes

334	531,441	276	Fourth or fifth
335	There are 29 solutions ranging from 5 silver dollars, 1 quarter, 2 dimes, 2 nickels and 7 pennies to 9 halves, 4 quarters, 2 nickels and 2 pennies.	277	3 3/4 quarts extra
		278	Barn near lower right corner
		279	Monday, February 7
		280	Eric wins (He is $\frac{1}{24}$ of an hour ahead.)
343	4,310,577 bulbs	281	827.5 minutes (13 hours, 47.5 min), if you count the 45 minutes of rest
344	$2,046,000	
352	$50	282	49 miles
353	33 meals	283	47 loads
361	Might have 1, 5 or 9 large cones	284	About 22 1/2 days
362	See page 194.	285	12 ways (2 ways go by by A and D, 2 by A and H, 2 by B and D, 1 by B and H, 1 by C and H, and 4 ways by E and H)
363	See page 196.	
171	1.70	
172	46 ft of paper 11-inch wide or 32 ft of 10-inch paper plus 14 ft of 11-inch paper.	
		286	495 people per hour
		287	2910 minutes or 48.5 hours
173	1655 hours	288	376 times (counting 1:00:00, etc., and all 24 hours)
174	100 ounces	
175	510 miles	371	156 pounds
176	One hour	372	60 miles an hour
177	7 packages	373	Yes. He will have 850 liters of oxygen to spare.
178	595 kwh	
179	1060 words/month	374	About 61¢/hour
180	52 books	375	28.86 miles/gallons
181	4 days total	376	144 bricks
182	8 quarts	377	75 feet
271	Lawrence	378	$10,888
272	16 columns × 8 rows = 128 photographs	379	108 ways
		380	.4 kilowatt-hour
273	1216 2/3 hours	381	2.27 foot radius
274	102 pounds	382	See page 282.
275	276 bricks		

SECTION I

Getting Started

C H A P T E R   1

The First Weeks

Preliminaries

Students enter school in September with a variety of attitudes, of which two are the most prominent. The first is that "I can't do word problems, but if you tell me what to do I'll do it . . ." The second, is that most of the students are relieved that summer is over, if for no other reason than because they have been deprived of social contacts. The goals of this section are to take advantage of this last attitude, the desire for social contact, and thereby to change the first. For these reasons, *all the activities in this section should be done concurrently.*

Our scheme is to have children work with each other and help each other in groups of four. These groups provide the students with the social contact they desire and reduce anxiety by creating an open and supportive environment. Creativity flourishes so that varied and imaginative solutions come to light in this environment.

The social setting of groups of four and the successes your students have in problem solving will meet the goals set for this section. While working in groups of four, your students will learn to ask precise and thought-provoking questions. They will become comfortable and better at explaining their thought processes. By the end of this period, respect for and cooperation with each other will become the usual class attitude.

Changing the mind-set activities will contribute greatly in eliminating their fearful and/or inadequate feelings. Students will enjoy the challenge and relish the success they have in solving problems. Moreover, they will view problem solving as something to use in their daily lives, not as something to be left behind in the classroom. When involved in the logic games, they will even ask for more!

You, as their teacher, will have opportunities to take a more active part in the class activities, also. Your interest and enthusiasm are contagious, and important because teacher support and encouragement are student motivators. There will even be times when you can be an active problem-solver yourself.

3

LEARNING
must be
EXPERIENCED

I hear, I forget.

I see, I remember.

I do, I understand.

The Teacher Becomes A Guide

The first eight weeks is a vital and challenging period for you, the teacher. Your customary role must change. You must step back and allow your students to have successes on their own. They need to talk to each other, explain to each other and listen to each other. (A quiet classroom, during a group-of-four activity, means no work is going on.) Sometimes their solutions won't work, and you will have to allow them to discover it themselves.

Nevertheless, teacher-led discussions are essential to the success of these activities and should be conducted in an open, accepting, non-judgmental way. One of your responsibilities is to model desired behavior and to lead your students into making wise decisions. This means most of them must first unlearn their negative, judgmental ways. When this has been accomplished, you will be able, by modeling, to help them make decisions that are beneficial to everyone.

Begin with setting up groups of four. Concentrate on this for the first two weeks. Use everyday procedures as group-of-four tasks to foster cooperation and the willingness to help others. For example, assign each group one day of the week to be responsible for presenting current events to the class. When passing out supplies or materials, you might also ask that each member of a group get enough for everyone in their group; i.e., one person gets the books, and another gets the papers. When finished, one person hands in the papers and another returns the books. You should also utilize the content task ideas presented in Beginning Activities, page 6.

As you and your students become comfortable with this mode of operation, start the Changing the Mindset Activities and the Logic Games. For the next six weeks, emphasize the advantages of groups of four and provide many opportunities for your students to work successfully in their groups.

We have defined a management system of three major rules which, when understood and followed by the students and teacher, will ensure effective interaction within the group of four.

GROUPS OF FOUR RULES

1. You are responsible for your own behavior.

2. You must be willing to help anyone in your group who asks.

3. You may not ask the teacher for help unless all four of you have the same question.

Although two weeks have been set aside for these rules to be discussed, practiced and understood, continued work on these should be on-going all year. Initially you will need to explain each rule separately, citing student examples to make them clear. As the year progresses, these rules, as the set of class-room management procedures, will need to be reviewed in their entirety.

The **first rule**, being responsible for your own behavior, is not unfamiliar to students. They can explain it easily and have usually heard it from every teacher they have known. However, only a few students seem able to accept it on a daily basis. Much class discussion on this rule is usually needed with many examples being given by both the students and the teacher. Respect for and considera-tion of others must also be discussed at this

time. Encourage your students to work out their own difficulties.

Students sometimes balk at **rule two.** It is usually strange to them and may cause con-flict to a highly competitive student. However, during class discussion you can turn the rule around. Point out that each student also can *receive* support from a group of three willing helpers. Interestingly, in actuality, this kind of cooperation improves self-reliance. Many students are willing to answer their own ques-tions and solve their own problems once the side benefit of getting the teacher's attention is removed.

Rule three needs the most practice, by the students and particularly by the teacher. It is difficult for the teacher to ignore a casual question, and many students have habits formed over several years. Thus, all concerned must exercise patience and practice. Learn to smile in response to such questions as "Is this the right paper?" and walk away. If this is too difficult at first, relay the question to the student's group of four. This rule is of great importance, however, as it provides a valuable framework for class interaction, with each other and/or the teacher.

Furniture And Random Grouping

There are three aspects to groups of four. The first is the management system already discussed; next is the physical arrangement of the students and the furniture; and the third is the random grouping method.

In a typical class, students sit at eight tables in groups of four. If there are groups of three and/or five, they sit at one of the round tables. If there are two students left—they join a table of four and work basically as two teams, two and four.

Alternatively, thirty-two desks can be arranged in eight groups of four, as pictured. You can see by the arrows representing lines of communication that this physical setting itself provides the opportunity for optimum interaction. No one is ever isolated. Three sets of two pairs can be communicating, discussing, explaining, arguing, clarifying and/or smiling simultaneously.

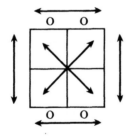

To achieve optimum interaction, students must be grouped randomly. The easiest way to do a random grouping is to number the tables or desk clusters, have the students draw a number and then have them sit at the correspondingly numbered furniture. The students should know that this seating arrangement is for a finite period of time. One week seems to be the ideal time period for fourth through sixth graders; seventh and eighth graders can usually operate for two or three weeks, depending on personalities and activities.

Generally students accept the outcome of this lottery. However, one or two students might attempt a trade or some other maneuver. This attempt is quickly abandoned when they hear the alternative—to have their seat selected by the teacher.

Beginning Activities

After a day or two of discussion on the rules for groups of four, launch the students into a review unit with activities that involve a partner from their groups of four. As both partnerships will be working on similar projects, all three rules can be practiced. It is essential that the unit be a review unit, for all the students' energy needs to be directed toward observing and practicing the three rules.

An interesting topic often used successfully for the first week of school is **graphing**. Prepare a list of questions of interest to the students, such as favorite color, speed of bicycle, favorite snack, Zodiac sign, distance of home from school, and others, allowing room for their additions. Then gather this data and distribute single topics to pairs of students. Their assignment is to organize the data and present the results in graph form. The graphs are then displayed around the room and can be used for a variety of follow-up activities.

Another successful beginning is to set up a unit of study on **measurement**. Possible measurement topics include non-standard, English or metric systems. These are logical topics because the students have some familiarity with them, and actual measuring is easier to do with two people than one. Sharing measurement materials and the discovery of inexact answers within the group will promote problem solving.

Either of these units of study takes about a week for completion. This provides the teacher with a short time each day to review the rules and clarify any confusion the students might have; it also allows the teacher time to move freely from group to group, interacting when appropriate. In the weeks ahead, projects should be presented that will necessitate cooperation among all four students.

C H A P T E R   2

Understanding Groups of Four

Optimum Interaction

INTERACTION IS AN INTEGRAL PART OF THE LEARNING PROCESS.

The best conditions for optimum interaction are obtained by clustering students in groups of four. This interaction takes place because of the continual open lines of communication. With groups of four, if any two members are communicating, the remaining two are not isolated. They may not be talking, but the option is available; they are not shut off.

If this number drops to three, someone inevitably gets left out of the interaction. The lines of communication are still here, going in every direction, but each person can only communicate with one other at a time. And while these two are communicating, the third member is isolated.

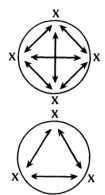

Avoiding Isolation

LEARNING DOES NOT TAKE PLACE IN ISOLATION.

It is accepted that learning does not take place in isolation. Any time learning occurs some kind of interaction and/or communication is needed. You probably know some students who talk out loud while studying. This interaction with their own thoughts and ideas is necessary and is often followed by a discussion with a fellow student or group of students. Students working in groups of four always have access to interaction because of their physical setting and the management rules.

Oral Language Development/Noise Level

ORAL LANGUAGE DEVELOPMENT PLAYS AN IMPORTANT ROLE IN LEARNING.

During groups of four interaction much language development takes place. Without the language that goes with the learning, very little learning will take place. Students need to be able to state what is happening, to talk about it and to ask questions about it. Through this verbalization they learn how to ask questions that will generate a response that is helpful to their understanding; they also learn how to explain their own reasoning processes.

With all of this verbal discovery going on in the classroom, what about noise level? Although there is a great deal of talking in the classroom, this type of interaction produces a pleasant noise, not a distracting or unpleasant one. This effect is achieved by making the students responsible for monitoring their own voices and topics up for discussion. Of course, this does not come about naturally. Many class discussions must be held on the topic, and the advantages that come with this interaction should be pointed out.

A Supportive Environment

LEARNING TAKES PLACE IN A SUPPORTIVE ENVIRONMENT.

An important aspect of the group of four is its supportive environment. Many students who would never pose a question in front of thirty peers are willing to ask it within their group of four. Thus the group encourages creative thinking, and each member feels safe to use trial and error methods. Anxiety is greatly reduced when an open and supportive environment is present. Furthermore, a supportive atmosphere provides an opportunity for the minority opinion to be heard.

Emphasizing the Process

EMPHASIS MUST BE ON THE PROCESS, NOT THE ANSWER.

Often a student's question will produce a question response and the group becomes actively involved in problem solving. This generates new ideas and new questions and new answers—it certainly gets students

thinking. Creativity flourishes and multiple solutions are proposed. When this takes place, it is important for us, as the teachers, not to interfere, and not to give answers. It is through understanding and emphasizing the process that our students will improve their thinking skills and techniques. Not all of their questions need answers! It is important that sometimes students make decisions and take actions without an answer key or answer person. This is called "preparing for the real world."

Student/Teacher Advantages

1. With some teacher guidance, a great many social things get taken care of; learning how to deal with people who are different, acquiring patience and tolerance, learning to be respectful of others, seeing a classmate in a new light—and developing self-reliance (not always looking to adults for solutions).

2. Within the groups, each student gets many opportunities to be both the teacher and student, since activities and groups change often.

3. Students realize that solving problems together is often more efficient and enjoyable than working by themselves.

4. This format reduces a teacher's class size, the typical class now consists of eight "students."

5. When raised hands do occur, there are four eager listeners waiting, and they will have a well-defined question.

6. Students and teacher alike get immediate feedback.

7. This management system allows the teacher to be a learner, too.

Four Roles for Interaction

In order to facilitate the groups of four, the teacher needs to be aware of the four roles necessary for optimum interaction. The four roles are the questioner, the summarizer, the prober and the doer. These roles are ever-changing, seldom being played by a participant for longer than five minutes. In order to better understand them, here are some typical comments made by each:

Questioner: "What did she say?" "I don't get it." "Where are the scissors?"

Doer: "I'll get the blocks, you get the markers." "Let's divide the task up into parts." "All right, everybody—let's get started."

Prober: "I wonder if this will produce a pattern." "If we used triangles instead of squares, we might . . . ." "I want to find out if this is true for all rectangles . . . ."

Summarizer: "Now let's review the directions." "Remember to put your name on your paper." "Oh, that's a nice design, John. Don't you think it is interesting, Maria?"

Your job as teacher is to rove around the classroom, listening. (Always listen to the group you are *not* looking at.) When it is obvious one of the roles is missing from the group of four, join that group and take the missing role. Someone else will soon take over this role again, and then you may move on to observing and participating with other groups. When joining a group, it is imperative that you be on their level—sit on a student chair, on the floor, or whatever; join as a member, not as the boss and/or the teacher.

Rules for Teachers

Once the groups are in action, you must continually heighten your awareness and improve your listening skills. There are four very important rules for teachers.

1. Do listen.
2. Do interact.
3. Don't ignore.
4. Don't interfere.

Your first responsibility as teacher is to listen carefully, continually and unobtrusively, remembering not to interfere as long as the group is interacting, with all four roles in play. These four roles must all be in operation, and it is your responsibility to see that they remain in action. Maintaining an interest and aware-

ness of what's happening and becoming a member of the group when necessary are essential, however. Students don't like being ignored; a smile here, a touch there, a nod across the room, a look behind or a quick stop to join in the fun add much to this supportive working environment. Students always seem to know the whereabouts of their teacher. The students react if a teacher spends too long talking with a visitor or working with one particular group.

Changing the Mindset of Students

Preliminaries

Many students today are accustomed to "being done to and for" in their daily lives. The television entertains them, their parents organize their social lives and leisure activities, and we, their teachers, cajolingly spoonfeed them knowledge, often even providing necessary tools (i.e., paper, pencil and books). All that is required of them is to show up—often at their own convenience.

It is our hope to reinstate in these young people a desire to solve problems, to find pleasure in a challenge and to seek for better solutions. In order to reach these goals, it is necessary to change the mindset of our students.

By the time students reach the intermediate grades, they have already formed certain opinions. The mindsets that need to be changed are exemplified by comments like these: "I hate math!", "Math is boring!", "I'm dumb in math.", "I don't care!", "Show me what to do!" The activities introduced in this chapter work to combat these attitudes. Some of

them are open-ended, giving the students the final say in their outcomes; others are totally success-oriented, enabling each student to succeed. Both classroom problem solving and curriculum problem solving are included. All of the activities involve the group of four, which provides a supportive environment. The goal of this section is to convince students that they *can* solve problems and that they can, in fact, actually enjoy doing so.

Classroom Problem-Solving Activities

There are many opportunities in routine classroom life for teachers to step back and allow the students to solve their own problems. This technique usually requires patience and determination because it is often easier for teachers to do it themselves. But in the long run, life in the classroom will run much more smoothly when students can deal with their own problems in a satisfying and acceptable (to you and them) way. Here is a typical example.

> Mary has lost her pencil; John's pencil lead is broken; Lupe left her pencil unattended at the table and it was stolen; Ed has three pencils, two of which he acquired in the last ten minutes.

If the teacher resolves this problem today, the odds are it will happen again tomorrow. It is better to give it to the class to solve, not on an individual basis but in their groups of four. There are two reasons for this. First, the students get practice working in their groups in a non-threatening situation. Second, this is a real problem, and in its solution a classroom procedure will evolve that everyone can live with for the rest of the year. To focus the discussion, the following questions might be written on the chalkboard.

> What shall we do about lost pencils?
> What shall we do about found pencils?
> When is a proper time for pencil sharpening?
> Are there any exceptions?

This activity may take up an entire math period. The time is well spent, however, if the students have gained a feeling of success in problem solving. A side benefit is the practice they receive in working in groups of four and coming to a consensus. A stipulation for problem solving of this type is that the group must agree on their solution before it can be presented to the class.

Several other problems students have solved are: entering and leaving the room (putting an end to pushing and shoving); getting supplies for the whole group from cupboards and shelves; sharing and rotating jobs such as clean-up, bulletin boards and attendance; and planning class parties (including collecting money, making a budget, planning a menu, decorations, arranging for entertainment, getting class approval and carrying out responsibilities).[1]

Students are generally eager and capable, when given the opportunity, to solve problems. The small group interaction (verbal) that takes place over playground disputes often provides rich material for in-class analysis. Frequently these disputes are amenable to group-of-four discussions, and either the students or the teacher may suggest that this occur. When this happens, each group is given the "hypothetical problem" and comes up with a solution; these solutions are then shared with the whole class. While solving the "hypothetical" problem, the students incidentally are solving the actual dispute. Here is an example.

> The bell rang, and as I met my class, it was obvious that things had not gone smoothly. Yet another game of Pogo had been broken up by "those two sixth grade boys." It was time to deal with this issue. My students took their places in their groups, and I gave them the problem.
>
> > Several members of our class take our ball and play Pogo at every recess. This was so, again today. About half-way through recess, Paul and Tom (fictitious names) ran through the game, kicking the ball out into the field. By the time the ball was retrieved, recess was over.

I told my students they would have about ten minutes for their teams to come up with solutions. At the end of the time (closer to fifteen minutes) I asked each group to report their conclusions to the class. The students offered the following recommendations.

Group 1

Tom and Paul are always in trouble. Nobody likes them. They should have to sit on a bench during recess.

Group 2

Tom and Paul are bullies and the teachers know it. The teacher on duty should watch them all the time.

---

[1]But . . . have you ever attended a party with 32 ten-year-olds where there were two gallons of cold punch and no cups? Luckily, someone in my class knew how to make a cup using paper folding—we had a great time!

Groups 3, 4, 5, 6 and 7 (They gave varying wording, same idea.)

Tom and Paul always wander around the first half of recess. Nobody likes them and they never get asked to play. They get bored and probably feel bad so they wreck other people's games. How about inviting them to play at the beginning of recess?

Group 8

Tom and Paul are mean and stupid—nobody likes them. If a big enough group got together, we could beat them up after school and teach them a lesson.

As solutions were presented, I allowed no comments. At the end of the reports, I asked for class reaction. The class, with some fear and trembling, decided to go with asking the boys to play. Two members were appointed as diplomats to offer the invitation. Several others volunteered to play in the game. We discussed what the rest of the class would do. "Watching" was discouraged by me.

Much to our surprise, the solution worked.[1] During the next few weeks, Tom and Paul played Pogo and were very cooperative. It also improved their reputation, and they began joining other games, also. As the year progressed, the initial incident was forgotten, even though Paul and Tom occasionally got into trouble.

Curriculum Problem-Solving Activities

While group-of-four practice, classroom problem-solving activities and logic games (see next chapter) are going on, it is necessary to present more opportunities for the students to practice problem solving within their

_____

[1] In fact, during the first few months of school my class became very good at talking through their own disputes. By February their behavior was generally cooperative and supportive. The students were willing to discuss their differences rather than act them out physically.

groups. Here are some examples of curriculum-oriented problems. Students should be given one problem of this type each week to work on in groups of four. The group structure is helpful to the individuals in motivating them and helping them find solutions.

I Consumerism: Which orange juice would you buy?

In all parts of this country there are six to ten varieties of orange juice on the grocer's shelves and in frozen food bins. There are frozen concentrates—some real, some imitation, some pure and some with additives; there are canned juices—different brands and contents; the powdered; the freshly squeezed and bottled; and oranges themselves! Give the students three categories for rating purposes; each team of four should come up with recommendations for purchase based on their own findings regarding the taste, nutritional value and price per serving.

II Art: Any display object or design may be used for this activity. The interaction in this task is derived from the problem solving needed to make or copy the object or design without any directions.

I have a stand-up black cat I made by cutting and folding a single piece of construction paper. Its tail and head are attached with a dab of glue. The eyes, mouth and whiskers are cut out of scraps and glued on. I put it on display . . . my students get one sheet of construction paper, scissors and glue, several scraps, if desired, and *no* directions. It is a very successful art project and problem-solving experience. (Some of my students end up with tiny kittens—that's O.K. If you cut opposite the fold instead of on the fold, you end up with two small kittens instead of one large cat.)

III Arithmetic

Have your students plan a holiday meal for their families. They are allowed three dollars per person. Their task is to make out a menu, to decide how much of everything they need and to go 'shopping' for it. They should keep track of the prices, brands, amounts, etc., as they may need to revise their plans. (This provides a nice homework activity.) The students must report back to their groups and they get some interesting feedback. Follow with a class discussion and let them make revisions. (Many students, for the first time, will decome aware of the price of food and the nuances, such as sale items, specials, and convenience foods.)

IV Pendulums (Science)

I purchased several pounds of metal washers at a local hardware store, ordered a ball of string from school supplies and obtained rulers from our class supply. After making and demonstrating a pendulum for my class, I told them they would also have this opportunity. Their task was to find out as much as possible about the workings of a pendulum. As each team of four made discoveries, they were to formulate each one into a question. These questions were posted for all students to see, thereby encouraging and motivating all teams not only to find out the answers, but to write questions of their own. These are some of the questions:

> Does the string length make a difference?
> Does the number of washers make a difference?
> Does the starting position make a difference?
> Are the number of swings consistent with a given number of washers?
> What about a push-start as opposed to a drop-start?

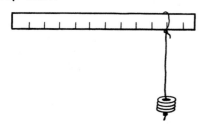

At this point, I felt a need to interject some suggestions. We needed some constants in order to discuss the findings of various teams. The class decided to time ten swings for consistency, excluding the first swing. For them,

a swing consisted of a forward and back movement. All timings were to be rounded to the nearest second.

The students quickly agreed upon their first set of questions, and groups began making discoveries and asking such further questions as:

> What effect does fish line have as opposed to string?
> Does a rounded fulcrum change the outcome any?
> What differences, if any, occur when various other weight materials are used? (clay, paper clips, tennis ball)
> Given the times at five swings and ten swings, can you predict fairly accurately for fifteen swings?
> How does a grandfather clock work?
> Could our pendulums be used as time-telling devices?

Keep in mind that my students came up with these questions, not me, and that the questions were formulated by the groups after they had discovered the answers.

V Decision-Making

At some point in early December, give each student a hypothetical fifty dollars and tell them to select a present for each member of their family. They must include the appropriate sales tax and may not go over, even a penny's worth.

When they cut out pictures of what they're purchasing, have them list the price, total all gifts, compute sales tax and show the total spent and money left over. (It helps to collect catalogues and flyers all fall.)

This task produces much interaction, advice seeking and giving, and for some, frustration in dealing with "which family?". The decisions they make and the discussions held are mind-opening and elicit heart-felt concern.

VI Giving and Following Directions (Communication)

Each person in the group of four needs an identical set of materials: six or seven building blocks, six or seven rods of different lengths and colors, or any other assortment of three-dimensional objects. (Each group does not have to have the same things, only the individual members within each group of four.) Also, each person must construct a barrier so that no one in the group can see his or her working space.

One of the four students builds a structure using some or all of their materials and then gives verbal directions so the others can build an identical structure. The goal is that when all the directions are given, and the barriers are lifted, all four structures will be exactly the same. A whole class discussion is held on the importance of the language used. Allow enough time for each member of the group to be the direction given.

Try this activity with no questions or comments allowed. Try it another time with the builders asking questions and receiving responses. How does this change the results? Try it a third time, allowing the student giving directions to move about and observe the other builders and to give directions accordingly. How does this change the results?

VII Pentominoes (Mathematics)

These geometric activities deal with the ideas of shape, size, symmetry, congruence and similarity that get students actively involved in problem solving with concrete materials. Each team of four is given twenty cardboard squares (five for each person) and the following directions.

Arrange five squares into different shapes following the rule that edges must always be completely touching.

This is acceptable:

This is not acceptable:

If your team finds two shapes that look different, but would fit exactly on each other if you cut them out and moved them

around, then those two shapes count as the same. They are congruent.

Record your team's work on squared paper. There are twelve different shapes possible. Try to find all of them. (With younger students, I do not tell how many shapes are possible; instead I ask how many they can find.)

Here are two follow-up activities:

1. Just by looking, try to decide which of the pentominoes would fold into a box. Circle them on your recording sheet, and put an "x" in the square you think would be the bottom of the box, opposite the open side. Then cut them out and check your predictions.

2. Someone in a factory bought lots of cardboard that was five squares by four squares. They figured that each sheet of cardboard could be cut into four pieces so each piece would fold into a topless box. How should the sheet be cut?

VIII Palindromes (Mathematics)

Here is a group-of-four activity that involves arithmetic. Each person on the team needs a 0-99 chart and a color key. Explain the task to the whole class and then have each team go about sharing the work load and finding the solution.

A palindrome is a number that reads the same frontwards and backwards, like 44, or 252, or 8118. Here's a number that is not a palindrome: 137. But you can change it into a palindrome with a bit of addition. The trick is: *Reverse the digits and add.*

It took one addition, so 137 is a one-step palindrome. Sometimes it takes a little longer to get a palindrome. Take 68, for example. It takes three additions, so it's a three-step palindrome.

```
 137       68
+731      +86
 868      154
         +451
          605
         +506
         1111
```

Investigate all the numbers from 0 to 99. On a 0-99 chart, color in all the numbers that are already palindromes in one color. (This will include all the one-digit numbers.) Then choose another color for one-step palindromes, another for two-step palindromes and so on. *Beware of 98 and 89!* The chart is on the next page.)

An excellent follow-up activity for this is to go on a word search for palindromes. A little competition among teams might even be applicable. After a large word list has been accumulated, teams could try to write palindromic sentences.

We hope these ideas have stimulated your own thoughts and that, as you proceed, you will add more of your own. You will find students to be once again curious and eager to meet new challenges. All they need is a willing and open guide.

Palindrome Recording Sheet

What colors are you using? Color in the small squares to show.

☐ already palindromes   ☐ 4-step
☐ 1-step   ☐ 5-step
☐ 2-step   ☐ 6-step
☐ 3-step   ☐

0	1	2	3	4	5	6	7	8	9
10	11	12	13	14	15	16	17	18	19
20	21	22	23	24	25	26	27	28	29
30	31	32	33	34	35	36	37	38	39
40	41	42	43	44	45	46	47	48	49
50	51	52	53	54	55	56	57	58	59
60	61	62	63	64	65	66	67	68	69
70	71	72	73	74	75	76	77	78	79
80	81	82	83	84	85	86	87	88	89
90	91	92	93	94	95	96	97	98	99

C H A P T E R 4

Thinking Logically

Preliminaries

The activities in this chapter teach reasoning through what students think are games. They are fun to play, develop reasoning skills of a high order, allow students to have positive and successful experiences as part of their groups and also begin the transition to more formal academic learning within the groups. These games all share the following features.

1. They give students a chance to reason, to make complicated, multi-step deductions, to evaluate the reasoning of others and to explain their own reasoning.

2. They focus on the questions "What do we know?" "How are we *sure* we know it?" and "How can we test a reasonable guess?"

3. They can be played in a group of almost any size—from two people up to a whole class.

From these games there are two important ideas which your students will hopefully learn.

The first is that they are capable of more involved reasoning than they had dreamed possible. The second is that there are such things as *general* strategies—approaches which will always work. The tasks here are to make them aware of their own thinking styles and to realize that alternative, possibly better, thinking styles exist. As the year progresses, they will start to replace their unconscious, *ad hoc*, set of thinking strategies with a consciously chosen, coherent set. But recognizing the existence of universal strategies is the first step.

There are three games or activities for you to use in conjunction with the group-of-four practice sessions and the Changing the Mindset problem-solving tasks. A proven successful format for using these games is for you to play the game with the whole class once or twice, then turn your students loose to play in their groups of four. The games, which will be presented below are the Digit-Place Game, Poison and the Color Square Game.

Usually the students will play so that three of them are trying to figure out what number

17

or design the fourth one has in mind, but sometimes they will pair up. The first way has the advantage that more cooperative learning can go on as the three discuss guesses. Keeping the sessions short (10 to 15 minutes) is an essential part of this activity. Here are two further guidelines for these activities:

1. Encourage your students to look for general strategies, *but*
2. DON'T tell them any of your own strategies.

It is quite possible for you to do some discrete modelling of your own approaches and hope they will catch on to better ways. But it is still important for them to reason out the ideas themselves. A parallel rule is to keep the games on their level. You can adjust the levels of difficulty up and down to suit the daily mood of the class.

Most children will move on to more difficult versions as they get better. There are always a few students, however, who seem to be afraid to try new versions of the games. Once they have mastered the strategy of one version, they resist the slightest change on the grounds that then they won't know what to do. By and large, these are the students whom we are most interested in reaching—those who think that strategies are to be memorized, not discovered and understood. You must prod these students to move on when the time is ripe.

After your students have realized that general strategies exist, it is time for them to begin developing good approaches to playing these games. Let them founder for a day or two with the complexities of each game before you either model good strategies or give them some hints where to look for strategies. As students become more proficient, homework assignments to emphasize particular aspects of any game are useful, both for the students and for their parents.

The Digit-Place Game

We first learned of this game from an Israeli friend who said his children liked to play it at school. Since then, a logically-equivalent version has been marketed using colors, words and now numbers under the name of "Master Mind". Whoever marketed the game has made a fortune. And with justification: it is a good game.

In our version, one person thinks of a two-digit number and the other players try to discover what the number is. They do so by naming any two-digit number. In return, they are told the number of correct digits and the number of correct places in their guess.

A sample game will probably clarify the idea better than any explanation. In this game the number to be guessed is 48. The first column represents the successive guesses while the second and third represent the number of digits and places correct. The first guess has no digits (and, hence, no places) correct because 48 has neither a 3 nor a 9 in it. The second guess, 47, has one digit correct (the 4) which is in the proper place. So one place is also correct. The third guess, 86, has one digit correct (the 8), but it is in the wrong place. Thus D = 1 and P = 0. The fourth guess, 67, has no digits and no places correct. The final guess is the correct one.

Guess	D	P
39	0	0
47	1	1
86	1	0
67	0	0
48	2	2

Some comments about the game are worth noting. First, it is much easier to play if the chosen number has different digits. In fact, that stipulation is usually made when the game is introduced. Second, only play until the person KNOWS what the number is, NOT until they actually state it as a guess. This distinction may seem too subtle to be important, but it has a great deal to do with how the children view the strategy of the game. Such a rule focuses attention on the notion that the purpose of the guesses is to gain information, not to hit haphazardly on the right answer. Such a philosophy also implies that the students must defend their answer by demonstrating the chain of logic that led to it. At the beginning, their explanations will be garbled and incomplete, but being inarticulate is largely the result of lack of practice. You will probably want to restate some of the logic yourself in better form or ask other members of the class to rephrase what has just been said until everyone understands. (Here is an opportunity to model *safe* judgmental behavior.) By the end of the year, your students should be reasonably adept at explaining their logic and being able to explain what you mean is almost as important as the logic itself.

To give the students an idea of how they are improving, have them keep track in a notebook of the number of guesses which they needed to get the number without pure luck (i.e., just guessing the number without a reason). At this juncture, the distinction between saying the number and knowing what it is will become important. In fact, you should probably have your students continue to keep records throughout all of the games which they will play.

The ideas which we have outlined above may be clearer if you read the following dialogue from an experienced teacher introducing the game to her class.

Teacher: "Today I want to teach you a new game called the Digit-Place Game. You will be trying to guess a number I am thinking of which is between 9 and 100. The number has two different digits in it—like 7 and 3 if the number I am thinking of is 73. Now the way we play is for someone—Bobby, you—think of some other two-digit number."

Bobby: "Uh. Is 74 O.K.?"

Teacher: "Certainly. Now I would tell you that your guess had one digit correct and it was in the right place: I would record it like this." (The teacher draws a diagram on the board similar to Figure 1.) "If Bobby had guessed 47, then I would have said there was one digit

correct, but zero places correct and recorded it like this. Sarah, give me another two-digit number."

Sarah: "77"

Teacher: "No. The digits can't be the same. Try again."

Sarah: "24"

Teacher: "Thank you. Here you have zero digits and zero places. After you get to playing the game for awhile you will find that is a very nice kind of guess to have. All right. Let's start a real game. I have a number in mind. Who is going to make the first guess? Maria?"

Maria: "19"

Teacher: (Making a table, see below) "$D=0$ and $P=0$. Rudolph?"

Rudolph: "73"

Teacher: "No. I'm not thinking of the same number as before, but that was clever of you. Here the answer is $D=1$, $P=0$."

Sally: "37"

Teacher: "Now $D=1$ and $P=1$. Consuela?"

We will give you the table below with the next several guesses. Basically no cleverness at all is used. This same general strategy appears at all levels—second-graders to adults. Don't fight it at this stage.

Guess	D	P
19	0	0
73	1	0
37	1	1
32	1	1
38	1	1
39	1	1
36	2	2

The teacher played one more game and left it for the day.

DAY 2 The first fifteen minutes were spent by having the students play this game at their sets in groups of four, each student having an opportunity to lead the game as time allowed. Then the teacher played a game against the whole class. At the end of the game she asked if there were any guesses that had been wasted or that the class could have guessed the answer to before she said anything. There was discussion about several of the guesses and the class concluded that two of the guesses had not been necessary at all.

DAY 3 The teacher began by playing a 3-digit game. The dialogue went as follows.

Teacher: "Today we are going to play the Digit-Place Game with three digits. That is the only change. All the other rules are the same. Right now I have a three-digit number in mind and I want you to guess it. Sam, give me a three-digit number for a first guess."

Sam: "293."

Teacher: "Thank you. No digits and no places. Now look. If we write all of the digits from 0 to 9 on the board; (the teacher does so) we can cross off the 2, the 3, and the 9. That helps us remember that none of those digits are in the number. Melinda, another guess."

Melinda: "356."

Teacher: "Here D = 1 and P = 1. What does that tell us?"

William: "One of those numbers is in your number and it is in the right place."

Teacher: "You are correct, William. So either the 3, the 5 or the 6 is in the right place. Do we know which one of the three numbers it is? Or do we know which one it isn't?"

Mario: "It can't be the 3, because we know the 3 isn't there. Sam already found that out."

Teacher: "How many of you agree with Mario? Hold up your hands. How many of you think he is wrong? No one? All right. Can you tell if the digit in my number is the 6 or the 5?" PAUSE . . . "Amelia. Give me a guess."

Amelia: "I'm going to try 296."

Teacher: "Here again D = 0 and P = 0. Now what do we know?"

Connie: "The 5 is in the right digit and it is in the middle."

Teacher: "What makes you believe this?" (Lets Connie explain.) "Class, do you agree with Connie?"

The teacher circles the 5 on the list of digits and writes a three-digit number on the board with a 5 in the tens' place and blanks for the units' and the hundreds' places. (See Figure 3.)

Teacher: "Do we know anything else?"

Melanie: "Yes. There is no 6 in the number, I don't think."

Teacher: "Why do you think there is no 6 in the number?"

Melanie: "Because before we said that there was either a 5 or a 6, but it was just one of them. And we just said that the right number is the 5. So the 6 is wrong."

Teacher: "Thank you for that explanation. Did everyone follow what Melanie said?"

After a pause, the teacher crosses the 6 off the list on the blackboard. At this point the blackboard appears as follows.

Guess	D	P
293	0	0
356	1	1
296	0	0

0 1 2 3 4 ⑤ 6 7 8 9   -5-

Note what the teacher is doing. She is helping her class make deductions by asking them what they know and why they believe their logic is correct. In addition, putting the extra information on the board makes the reasoning simpler for the students. Later on, particularly in the third game, putting down the intermediate steps will be of great value. Moreover, the whole notion carries over to solving word problems in a natural way.

Teacher: "Next guess. Larry?"

Larry: "154."

Teacher: "Here D = 2 and P = 1. What does that tell you?"

Larry: "Both 1 and 4 are in the number."

Tony: "No! *Either* the 1 or the 4 is in the number because we know what the 5 is already there and it counts for one of the digits."

Diana: "We also know that whichever is the right number, 1 or 4, it is in the wrong place because the 5 is in the *right* place."

After a brief discussion within each group of four, most students agreed with Tony and Diana.

Teacher: "I see most of you agree with Tony and Diana, but we still had better check. Warren, can you make a guess that will help us?"

Warren: "156."

Teacher: "Here D = 1 and P = 1. Now what do we know?"

Yoshi: "The 5 is the only right digit in that guess. So the 1 is wrong and the 6 is wrong."

Teacher: "Does everyone agree with Yoshi?"

After a pause, when most students have indicated agreement, the teacher crosses the 1 off the list.

Teacher: "Do we know anything else?"

Sean: "Yes. Since the 1 is wrong, the 4 has to be right up above."

Ann: "And we also know that the 4 is in the wrong place, so it has to go in the hundreds' place. So your number is four hundred fifty something."

The teacher circles the 4 on the list and puts a 4 in the hundreds' place of the three-digit number.

Teacher: "Do we know what the last digit isn't?"

Carol: "Yes. We know it isn't 1 or 2 or 3 or 4 or 5 or 6 or 9. So all it can be is 0 or 7 or 8."

Teacher: "Do we know which?" PAUSE. "Jill, make a guess."

Jill: "450"[1]

Teacher: "Here D = 2 and P = 2. Does that tell anything?"

Rodrigo: "It just says that 0 isn't in the number."

The teacher crosses 0 off of the list.

Teacher: "Ellen, make a guess."

Ellen: "457"

Teacher: "Again D = 2 and P = 2. What does that tell you?"

Maria: "I know the number. It is 458."

Teacher: "Why do you think 458 is the right number?"

Maria: "Because it had to be 457 or 458. And we just found out that it wasn't 457."

Teacher: "Does anyone else have a comment or question?" PAUSE. "This game took you seven guesses. Now you may play the three-digit game in your groups."

At first glance, it may seem unlikely at best that your fifth-graders, who turn into gibbering idiots at the sight of a simple word problem, are capable of such reasoning. Not so. In the security of a whole class, with no time constraint from you, they are able to pick their way through some fairly involved logic. Give your students 20 to 30 seconds to reflect on whether or not they can make a deduction before moving on. If a wrong analysis is made and no one catches it, just act as though the reasoning is right and carry on. When the class thinks they have come to the end of the game, it is an excellent chance for them to look back over their reasoning to see where they went wrong. If everyone agrees that the final answer is correct and it isn't, simply tell them that they have come to the wrong answer and that they have some more guesses to make. Then, when they finally do get to the right answer, go back over the whole sequence of guesses so that they can find out where they went wrong.

It is important that the whole class play the game against you (or one of the students) several times so that the weaker students can profit by the modelling of the better ones.

---

[1] While it is too early for the students to have developed such sophisticated strategies, it seems worth pointing out that if Jill had made the (non-obvious) guess of 750, she could have pinned down the number precisely by the answer. For if D = 1 in this case, then the units' digit is 8. If D = 2 and P = 1, then the units digit must be 7, while if D = 2 and P = 2 then the units' digit must be 0.

Another way to encourage discussion within the groups of four about what inferences can be drawn from various guesses is to play part of a game—say three or four guesses—with no class discussion at all. Then give the groups a minute to decide on what next guess they would make. When a group is ready, circulate and respond to each group privately.

Remember to keep the focus on what they know they can deduce from the answers on the board, but don't tell them the answers. Stress that they should be able to explain their answers as well, but expect many of them to know the answer without being fully conscious of their reasoning process. However, to develop this consciousness is one of our goals.

Let your class play this game for 15 to 20 minutes every other day for a few weeks, both within their groups and as part of the whole class; then move on to the second game, "Poison." You may also give your students a subtle assist in organized questioning strategies by assigning a homework assignment similar to the one below.[1]

Find the number.

Guess	D	P
123	1	0
234	1	0
345	1	1
456	0	0
567	1	1
678	1	0
789	1	0

Poison

This next game is probably familiar to both you and your class under one of a variety of names. We include it next because it has an optimal strategy which can be deduced and

because it leads naturally into one of the major problem-solving strategies: finding subproblems.

The initial game is very simple. Two players have ten blocks in the space between them. They alternately take one or two blocks from the pile until there are none left. The person who takes the last block, the "poison" one, loses.

Students quickly get the idea that whoever reaches 1 at the end of their turn is going to win. It takes them a while, however, to see that whoever can get to 4 at the end of their turn is also going to win. It takes them a very long time to see that 7 is also a winning position to strive for. DON'T TELL THEM. LET YOUR STUDENTS DISCOVER THIS FACT FOR THEMSELVES.

Here is a typical schedule for introducing the game, assuming that it is played about twice a week. On the first day, introduce the game briefly and play against a volunteer student. Then let them play the game in pairs in their groups. IT IS VERY IMPORTANT that the students play the game with objects—blocks, beans or even pieces of paper. Most of them are still reasoning at a stage where the objects are very useful in helping them think.

On the second day, have your students play in pairs again, but this time in a group of four with one pair playing against the other pair. Have the members of the pair discuss their moves with each other (there will be a lot of whispering) and suggest that they actively look for a way to win. This is a good day for you to circulate through the classroom and allow anyone to attempt to "poison" you.

On the third day, have them go back to playing in pairs (one against one within their group of four); this time, however, they should begin with seven blocks in their pile. The point of this change (although you should not tell them) is that they may find it easier to identify 7 as a winning position. Then they can work backwards to find the winning strategy for ten blocks.

On the fourth day, play against the whole class, this time using squares drawn on the

---

[1]The answer is 307.

blackboard. The first time, play with seven squares, then 10 squares and then twelve squares. Let some free discussion go on about how many squares the class representative should take. Try to push anyone who believes he or she knows what is going on to articulate their ideas as clearly as possible.

By this time your students should be ready for a shift from the concrete to the abstract. They have already been primed by playing the game with squares on the blackboard. Now have your students play the following game in pairs. One player chooses either 1 or 2 and subtracts it from 10. The other player subtracts either 1 or 2 from the remainder, etc. The one who reaches 0 is the loser. Then ask them if this game is like any games they have played so far. Most will recognize it as the first one.

While this is going on, hand out a few homework problems of the following sort: "You are playing 'Poison' and may take one or two blocks from the pile. There are six blocks left. How many should you take to be sure you will win?" "You are playing 'Poison' and may take one or two blocks from the pile. There are 29 cubes. Do you want to play first or second? How many blocks will you take to be sure you will win?"

At the end of this whole series of games, take a little time to point out that the three versions of 'Poison' which they have seen—using blocks, squares on the board and numbers—are all equivalent. This is also a good time to suggest that many problems can be made simpler by using blocks, pictures or other concrete objects rather than just numbers. They should not be afraid to change how the problem is represented, so long as they see the connection.

For the remainder of the year, students may play various versions of "Poison" in either concrete or abstract forms. For example, the original pile might contain 24 blocks and each player may take 1, 2, 3 or 4 blocks at a turn. Continue to encourage them to seek strategies. Finally, you might suggest that 'Poison' is a good game for playing with their parents or siblings especially on a car trip. They will probably leap at the opportunity to beat their parents at a game.

The Color Square Game

This third game in our sequence involves deductive reasoning in much the same way as the Digit-Place Game. The reasoning tends to be somewhat more spectacular in this game because a few well-chosen guesses can unravel the complete pattern. So that you will have some idea of what we are talking about, let us describe the game.

The initial game is played on a square grid, with three squares on a side. Each of the nine squares is colored; there are three red squares, three green squares and three blue squares. However, all of the squares of one color are linked together along edges. The diagram at right gives you some of the permitted and forbidden arrangements.

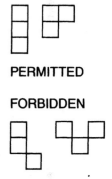

PERMITTED

FORBIDDEN

One person invents an acceptable pattern, involving the three blue, three green and three red squares, but keeps it hidden. The other players, using a blank three-by-three grid, try to reconstruct the pattern. In order to get information, they may request to know the total number of squares of each color in any row or column. The game is scored on the basis of the number of requests made before the players can accurately reconstruct the originator's pattern. As usual, the role of the teacher is to focus the students' attention on "What do you know?" and "What do you need to find out?"

What follows are two actual games, with commentary in order to better convey the flavor of the game. The first is an initial game played by a class. The second is a four-by-four game (played with four connected squares each of blue, green, red and yellow) by the same class at the end of the week.

Game 1:

Teacher: "Now that you have heard the rules for this game, let's play it." (The teacher draws a grid on the blackboard.) "I have a pattern in mind. Jaime, what row or column do you want to know about?"

Jaime: "What is in the top row?"

Teacher: "Two blues and one red. Remember that the blues don't necessarily come first; I just say them first." (At the right of the row the teacher writes "2B,R.") "Susan?"

Susan: "What is in the right, uh, vertical row?"

Teacher: "Right column. There are two blues and one green." (Writes "2B, G" above the right-hand column. Now the blackboard looks like the one below.) "Do we know any of the squares yet?"

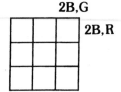

Gillian: "Yes. The one up there is blue."

Teacher: "Which one up there?"

Gillian: Top right-hand corner."

Teacher: "Why do you think so?"

Gillian: "Because . . . well, it can't be red or else there would be a red in that column and there isn't. And it can't be green or else the top row would say there was a green in it and it doesn't. So it's got to be blue."

Teacher: "So you are saying that the top right-hand square is either blue or red because those are the only colors in the top row. And the same square must be blue or green because those are the only colors in that right column. Therefore, the only color that lies in both the top row and the right column is blue. Does everyone agree with Gillian?" (After a pause, the teacher puts a B in the appropriate square.)

Note how the teacher paraphrased the student's words in a somewhat clearer form. The students can reason, but they generally lack the vocabulary to express their ideas clearly.

This kind of clear expression must be modelled extensively by the teacher.

Teacher: "Do we know anything else?"

Arnold: "I think that the one in the middle of the top must be blue, too.

Teacher: "Why?"

Arnold: "Because if the other blue were clear across in the other corner, the squares wouldn't connect like you said they had to."

Teacher: "Who agrees with Arnold?" (Most of the class agrees.) "Let me try to say what Arnold said in another way. Since there are two blues in the top row and we already know that the one on the right side is blue, the other blue either is in the middle or on the left side. But it can't be on the left side, because then the three blues don't connect properly. So the second blue must be in the middle." (The teacher puts a B on the top middle.) "Do we know any more squares?"

Francis: "Well, the other one at the left end up there has to be red since that is the only square left and we know that the top row has one red in it."

Teacher: (Puts an R in the top left-hand corner.) "All right. Do we know any more squares?"

Jennifer: "Yes. If you look at the right column, it's the same. The middle one has to be blue and the bottom one has to be green. It's just sort of like the top."

Teacher: "Like this?" (The other two squares are now filled in. The grid now looks like the one below.) "We've done pretty well with five squares filled in after only two guesses. Can anyone fill in any more squares?" PAUSE "O.K. Sam, make a request."

R	B	B
		B
		G

If the students were better players, they would realize that the square below the red must also be red; otherwise, the corner red is isolated. But it is very rare for someone in a class to see that implication the first time the game is played. After five or six games, you, as teacher, may push the issue by saying something like, "There is at least one more square you can fill in. Discuss it in your groups for 60 seconds, and then let's see if anyone knows."

Sam: "The bottom row."

Teacher: "You will like this. The bottom row has three greens. Can anyone fill in some more squares?"

Jerry: "Yes. Put Gs in each of the squares on the bottom row."

Teacher: "I agree. Any more?"

Matt: "Yes. The other two squares have to be the other reds."

Teacher: "You filled in all the squares with three guesses. Did you see how each time you figured something out it helped you figure something else out?"

Ask your students to play this game in their groups of four once or twice. You need to verify that the students understand the rules. Then go immediately to a four-by-four game.

Game 2:

(Remember this class has played the four-by-four game several times.)

Teacher: "This is a four-by-four game. There are four blues, four greens, four reds and four yellows. Francis?"

Francis: "What is in the top row?"

Teacher: "There are two blues and two reds." (Writes "2B, 2R" at the end of the row.) "What squares can you fill in with that information?" PAUSE "No one? . . . Kristie?"

Kristie: "What is in the left-most column?"

Teacher: "There are three blues and one yellow. Now do we know anything?" (Writes "3B, Y" above the column.)

Maria: "Yes. We know there is a blue in the top left corner and there are two more blues below it."

Teacher: (Writes B's in the appropriate squares.) "All right. Any more?"

Seldon: "Yes. There is a yellow at the bottom of that column, and there has to be a yellow in the square next to that one or else it won't stick to anything."

Teacher: "Does everyone agree with Seldon?" PAUSE "The yellow in the lower left-hand corner needs another yellow next to it. Any more?"

Marsha: "The top has to have blue, blue, red, red; like that."

Teacher: "Does everyone agree with Marsha!" PAUSE "All right." (Writes in "B, R, R," as requested. The grid now looks like the one below.) "Any more?"

3B, Y

B	B	R	R	2B, 2R
B				
B				
Y	Y			

Sandra: "Yes. The two squares below the reds must be the other two reds."

Teacher: "Does everyone agree with Sandra?" PAUSE

Don't worry. Someone will always disagree. But don't just ask for confirmation when the student is wrong. The students quickly pick up on this.

Teacher: "Well, if there is that much disagreement, let's wait until after another guess. Dennis?"

Dennis: "What is the second row up from the bottom?"

Teacher: "That row has one blue (which you already knew about), one green, and two yellows. Now do you know any more?"

Harry: "The yellows have got to be stuck together and stick to the ones we have. So they can't be on the far side; they have to be next to the blue, and that leaves the green over on the right."

Teacher: "Did everyone follow what Harry said?" PAUSE "He said that the two yellows in that row have to be side by side. And if they are going to touch the yellows which are already marked, the new ones have to be next to the blue square. Otherwise, the yellows won't be in a proper pattern. Then that leaves the green at right. Does everyone agree with this?" (After a silent PAUSE, the appropriate squares are filled in. The grid now looks like the one below.) "Any more squares that we know?"

Jennifer: "Those two squares at the bottom have to be green because only greens and yellows are next to them and we've used up all of our yellows. So they have to be green."

Teacher: "Does everyone agree with Jennifer?" PAUSE (The indicated squares are filled in.) "Anything else?"

3B, Y

B	B	R	R	2B, 2R
B				
B	Y	Y	G	B, G, 2Y
Y	Y			

Jaime: "Doesn't the square above the green have to be green, too? I mean, that is the only square that can connect to those other three greens."

Teacher: "So you are saying that the last green must connect to the other three. Do we know enough to fill in those last two squares?"

Lawrence: "They have to be red. That's all that is left."

Teacher: "Three guesses. You are getting very good at this game."

This game may be made easier or harder by adjusting the size of the grid; larger grids are harder. The four-by-four game seems to be the best size for most students. Some of the brighter ones want to try larger grid sizes, or sometimes they try rectangles instead of squares. You may also give your students the option of choosing different numbers of colors to put inside the grid. Generally speaking, the fewer the colors, the more difficult the game. (We know that sounds paradoxical, but try a four-by-four game with only two colors and you will see.)

A significantly different and deeper game may be made by allowing the students to choose any subset of squares on the grid for their guess. The one who already knows the grid color scheme then responds with the total number of squares of each color in that subset of guesses. Very good students may then investigate what strategies are good for dealing with this new game.

SECTION II

Understanding the Problem

Preliminaries

THE FIRST STEP IN SOLVING ANY PROBLEM IS TO UNDERSTAND IT.

The statement above would seem so obvious as not to be worth making, were it not that the failure to observe it is the most common reason students fail to solve problems. When faced with a word problem, in particular, some students may be so terrified of the whole idea of word problems that they fail even to read it through. Or they may simply skim to find a few numbers which they can add or multiply, depending on which section of the book they happen to be in, regardless of whether that operation makes any sense or not.

Why is it that perfectly normal, intelligent students consistently fall apart when given word problems? How can caring knowledge-able teachers continually be confronted with the "I don't get it" comments after thorough and clear explanations? Why is it that our teaching techniques that work so well on other topics fail to elicit much progress here? It seems to be because word problems involve a different set of thinking skills and must be approached in a different way.

By this time your students have had ample practice working in groups, and have successfully done a lot of problem-solving of various kinds. The cooperative spirit should be abroad in the classroom, and your students should now be ready to take their first step in solving word problems.

In this section, students learn how to read and understand word problems by writing problems for each other to solve. They will be led from the simple problems with which they invariably begin to complex problem formulations with lots of extraneous data, with the question hidden somewhere in the middle of the statement, and with solutions requiring more than one step. The simplest problem-solving technique—data organization—is introduced here in the form of a game. In addition, students learn to check their solutions against the information given in the problem to see whether or not they have the right answer to the right question. In short, by the time your students finish this section, they should be able to *do* easy problems and *understand* hard ones. Solving more difficult problems is the subject of the next section.

Teaching word problems this way takes longer than the usual approach, but the pay-off in additional understanding is well worth the time. In addition, the students practice and develop writing skills, so these activities can also be a part of your own English curriculum or coordinated with the English teacher's to everyone's advantage.

As before and throughout, the emphasis here is on students helping each other and striving for understanding. With the completion of this section, the following class problem-solving chart, *Understanding the Problem*, should evolve.

I.　Restate the problem.

II.　Clarify the question (includes the hidden question).

III.　Organize the information:
　　1.　Get rid of unneeded information.
　　2.　Find needed and assumed information.

IV.　Check the solution with the problem.

V.　Evaluate the solution and rethink the methods when necessary.

C H A P T E R 5

Basic Headlines
and Stories

Introduction

This unit of study takes about four weeks of
daily fifteen-minute sessions. There are three
major skills to be developed:

Restate the problem (story) in different
words.

Clarify the question.

Check the solution with the problem.

Here is a list of sequenced activities that will
develop these skills; these activities are
described in detail in the rest of the chapter.

1. Definition of headline and story

2. Teacher presentes headline; groups
   write stories (Do twice.)

3. {
   Student writes headline and story
   Groups of four work as editing
   groups
   Student rewrites within editing
   groups (Do these several times.)

   Student writes story; group trades
   papers for headline

4. Teacher leads discussion concern-
   ing the question
   Student rewrites using question

5. Student writes stories with questions

6. Teacher leads discussion concern-
   ing headline form

7. Student writes stories with questions

8. Teacher leads discussion concern-
   ing testing the solution

9. Student writes stories

10. Student stories as classwork/
    homework

Suggested Classroom Sequence

The first activity begins with a class discussion to define "headline." A front page or two
from a newspaper will greatly benefit this discussion. As teacher, your role is to guide the
discussion in such a way as to suggest the
definition of headline as a basic, factual summary of a story and/or article. Then, give the

students the following headline: $8+4=12$. Each group must now write a story to go with this headline. For example, a typical story would be:

$$8 + 4 = 12$$

John's little brother David is eight years old. David is four years younger than John.

At this point resist the urge to give any further clues and/or information. Also, do not mention incorporating questions in the story now; the stories should, on this first day, be simple narratives, nothing more. There will be quite a bit of talking in the groups. Keep in mind that any group having a real problem has the option to ask for help, provided all four members of the group have the same question.

The teacher's role during the story writing is to circulate among the groups, interacting when appropriate and collecting the stories. Read these stories privately and plan comments for the following day.

The next day read each of the stories aloud to the class. After each story is read, each student must cast a yes or no vote in answer to this question: do the headline and the story match? After the voting, you should make encouraging suggestions for improvements. (It is not unusual to have half of the stories receive a no vote the first time around.) This is followed by introducing another headline, using small numbers and either subtraction or multiplication like $4 \times 2 = 8$ or $11 - 6 = 5$. Each group must again write a story to go with the new headline.

The voting procedure is repeated, probably on the third day. During class discussion, the students should now begin to offer ways to improve the stories themselves. At the conclusion of this discussion, each student has to write a headline and story of their own. A short work period follows.

Begin the next class period by leading a discussion on a variety of writing skills: writing on a topic (headline), clarity, complete sentences: and paragraph form. Following dis-

cussion, the students take out their individual stories from the day before. The stories are to be read silently by the authors and then by fellow team members, so that corrections and additions may be made where needed. These editing groups provide much opportunity for the students to practice and improve their skills. Each team selects an example, in corrected form, to be read to the class and posted on the Headline bulletin board.

A review of the writing skills needed to produce a good paragraph begins the next class session. The students are then told to write a headline of their own (an equation), followed by a story to go with it. This is, in actuality, a three-day project. The first day is spent in writing the headline and story, with much self-correcting and indiscriminate group support. The second day involves formal group editing followed by a guided class discussion. Similar problems recur. For example, discrepancies may arise between the headline and the story. Is it possible to have an incorrect headline? If so, what should be done? (The author may change the story or the headline.) Is additional information needed? Decisions needed to be made, and the students must determine whether to change the story, or the headline or both.

On the third day of this activity it is time for rewriting, including final group editing and oral group-of-four sharing. That is, each author reads his or her own story aloud to the group of four. Each team then selects the order of oral presentations and shares these with the entire class, as time and/or interest permits. The students then turn these papers in for your perusal.

The next concept to be developed begins with an element of a guessing game. Ask each student to write a story on one side of the paper and to write its headline on the back, being certain to keep the headline a secret. The students then exchange stories within their groups, and each student writes a headline to go with the story he or she receives. The papers are then returned to their owners and results are discussed within each group. You then offer a summary through class discussion, guiding the students to the following conclusions.

1. Sequence within the paragraph form is important.

2. Clarity and completeness of each idea is important.

3. It is possible for one story to produce multiple headlines, each of which is correct. For example:

> John went to the bookstore. He bought two books. One of them cost $9.95, and the other one cost $14.95.

Here are two correct headlines:

$$\$9.95 + \$14.95 = \$24.90 \;;$$

$$\$14.95 - \$9.95 = \$5.00.$$

4. Asking a question is the way to get a story to produce one particular headline. This conclusion is the most important concept to be understood by the students, for it is essential to all problem solving: what *is* the question?

Students now rewrite their stories to include a question, which will not only produce just one headline, but the one particular headline they have in mind. These stories are tested in the groups of four and kept for ready reference for the rest of the week.

The next few days students practice writing stories that will generate specific headlines. These are continually shared, tested, and/or edited in the groups of four. This is where you will be kept very busy listening, interacting and teaching, all the time abiding by the third rule (no questions to the teacher unless all four group members have the same question) and maintaining the interaction within the groups.

Here is part of a group of four discussion which took place at this stage of the process.

Jon: "Has everyone finished?"

Doris: "Just a minute . . . "

Dave: "I'll read mine first!"

Jenny: "Does anybody have an eraser?"

Doris: "O.K., I'm done."

Dave: "My team won last night. I scored two touchdowns. Joe scored one and Shaun made one point kicking. What was the score?"

Jon: "Let's go through the steps . . . "

Doris: "Dave played football last night and wants to know the score."

Jenny: "Dave should know the score—he played in the game—besides, what about the other team?"

Jon: "We need to restate the question. Dave said "What was the score?" I think he meant— "How many points did Dave's team score?"

Doris: "That makes sense—Dave, is that what you meant or did you mean the score for the game?"

Dave: "Well, I meant my team's score. I guess I could have told about the other team's touchdown and then asked my question."

Jenny: "I think it's good both ways. Why don't we write both of the headlines and see which looks better."

Jon: "We need to look at the numbers we have."

Doris: "Let's start with figuring out Dave's team score. Is a touchdown worth six points?"

Jenny: Yes, everybody knows that. I wrote $12 + 6 + 1 = 19$."

Jon: "I wrote $6 + 6 + 6 + 1 = 19$."

Dave: "I wrote $6 + 6 + 1 + 6 = 19$. I guess that's because I know when Shaun made the kick."

Doris: "Mine is the same as Jenny's. They're close enough."

Jon: "The other team scored one touchdown—did they get the extra point?"

Dave: "No, they tried to run it and didn't make it."

Doris: "O.K., I have two headlines. The first one is $12 + 6 + 1 = 19$ and then 6 which really looks dumb."

Dave: "I like the first one by itself, I'm going to change my question."

Jenny: "You could include the other team's score and ask how many points you won by— then your headline would be super long . . . $12 + 6 + 1 - 6 = 13$."

Dave: "I like it better my way."

Doris: "I want to read my problem, now."

During this time, much emphasis is placed on question writing. The students become aware that some questions, no matter how carefully worded, can produce different headlines. As the guide in this process, it's up to you as the teacher to help them see that some headlines contain different operations but produce the same outcome. For example:

$$4\frac{1}{2} + 4\frac{1}{2} + 4\frac{1}{2} = 13\frac{1}{2}$$
$$3 \times 4\frac{1}{2} = 13\frac{1}{2}$$

As the students gain proficiency in story-writing, their stories may require more than one headline to answer the question. Some students do preliminary headlines in their heads, while others write out each one. Order in the headline now takes on a double significance, as it is needed to understand the question and it is singularly important within the various operations.

> Before going on vacation, Jack borrowed four books from a friend. He went to the library twice and checked out three books each time. How many books did he take on vacation?

$$4 + 3 \times 2 = 14 \text{ or } 4 + (3 \times 2) = 10$$

Many of the student stories should be collected and typed on dittoes. These can be distributed for classwork and/or homework. The students enjoy seeing their writing in print. You may want to let the students type their own. This technique is very helpful when conducting a whole class discussion, for all students will have before them the problem being discussed.

The class should now begin keeping a class problem-solving chart to serve as a review and also as a helpful aid in solving problems. The biggest gains the students achieve during this time are acquiring the skills of restating the problem and understanding the question. One of the advantages they have is that the author is nearby, and therefore any unclear thought can be redefined at the asking. Stu-

dents learn how to restate their own stories and questions and also those of their peers. The emphasis begins to shift away from an exact headline to a particular set of outcomes or a single outcome.

The students will frequently double check the solution with the problem (story), and you must capitalize on this. Lead several short class discussions, using student examples (with their permission) to point out the importance of testing the solution (headline) back with the problem (story).

The class problem-solving chart should now have three skills listed.

Understanding the Problem

Restate the problem.

Clarify the question.

Check the solution with the problem.

C H A P T E R 6

Silent Board Games

Introduction

Silent board games meet a number of needs. From their work with basic headlines and stories, your students should now understand the concept of restating the problem, clarifying the question and checking the answer. In general, they are least excited about checking the answer. Silent board games reinforce this concept in a dramatic way, for frequently in these games, the students' first guesses are not right and they *must* check. Moreover, the answers (rules) of the silent board games are most easily discovered when the data are organized in a coherent way. Organizing the data is the simplest, and perhaps the most effective, problem-solving technique. Through silent board games, then, students learn and practice a skill which is useful not only in mathematics, but in English and in all the sciences. It is another of the fundamental intellectual tools which everyone must have.

Silent board games fortuitously fall during December which is a hectic month in most schools. Since they are relatively quick games to play, they can be used as a fill-in whenever you have a few extra minutes. It is best to play this game against the whole class on a daily basis, interspersed with more work on basic headlines for the entire month. Begin with one-input games and stay with them until students have the idea of organizing the information for them.

Then, after discussing the principles involved, move on to two-input games. It is fun to let students take turns making up the games; just be sure that their rules are not too hard.

Activity

The games are easy to play. In the simplest version, the leader thinks of a rule for converting one number (the "input") to another (the "answer")[1]. The other members of the class attempt to guess the rule by suggesting input numbers and then looking at the corresponding answers. Usually the class sug-

---

[1] Mathematically sophisticated parents will be delighted to find that their children are studying functions.

gests about ten numbers, which are put in a column for input numbers, before the leader writes down any answers. After three answers are written down, any member of the class may come to the blackboard and write down a proposed answer beside any listed input number. If the guess is correct, it is left. If the guess is wrong, everyone gets a chance to look at it, and then it is erased.

All of the guessing of answers is done *silently*. No one needs to say anything. If the guess is wrong, just shake your head sadly as you erase. Frequently students will want to blurt out their guess of your rule, and the first few times that the game is played you will probably have to be firm with your class about the need for silence.

Here is the sequence:

1. Think of a rule.
2. Get ten 'input' numbers from the class.
3. Write down 'answers' to the first three.
4. Offer chalk to those who think they know the answer.

A typical early game is shown in Figure 1. This game is so easy that virtually every student will see the rule when you have finished writing the first three answers on the board. Feeling successful is a good way to begin.[2]

Input	Answer
4	48
2	24
7	84
100	
93	
6	
0	
5	
11	
1	

Figure 1.

[2]If you wish to emphasize the importance of checking you might put down Fig. 1 again at a later time where the first two answers are the same, but the third answer is 114. Here if N is the input number, the answer is $2N^2 + 16$.

Figure 2 represents a more complicated rule which will take a little longer, while Figure 3 shows a rule which will challenge all but the most able students. It is important to vary the difficulty of the game. After your students have dealt with a particularly difficult one, present them with a simpler one the next day.

Input	Answer	Input	Answer
5	13	9	90
10	28	3	12
2	4	6	42
100	298	1	2
7	19	5	30
4	10	12	156
5		20	420
1		10	110
9		2	6
14		4	20

Figure 2   Figure 3

In general, you should seldom discuss the rules which ink one number to another.* Sometimes at the end of the session, ask those students who know "your rule" to explain to the class *how* they know. The emphasis should always be on the thinking process involved, not on a single solution. Not all students can do all the games each day. That is fine, and be sure that your students know it is fine. However, the games should be chosen in such a way that every student can do some games each week.

After you have played this game with your class for three or four days, it is time to shift the emphasis to focus on good patterns of input questions. Here is how an experienced teacher did it after a silent board game which had resulted in the numbers of Figure 4.

---

*Possible rule explanations:
1. $3N - 2$; 3 times the number, take away 2; $3 \times - 2 = y$; 2 × the input + the input − 2.
2. $(N \times N) + N$; the number times itself, plus the number; the number times one more than the number; $N \times (N + 1)$; $N^2 + N$.

Teacher: "Today I want to play our same silent board game again even though most of you already know my rule. I am not going to do anything except rewrite the inputs in a different order."

The teacher creates Figure 5. Note the gap between 5 and 8.

Input	Answer		Input	Answer
9	42		1	10
5	26		2 ·	14
3	18		3	18
4	22		4	22
10	46		5	26
1	10			
8	38		8	38
25	106		9	42
2	14		10	46
19	62			

Figure 4   Figure 5

Teacher: "Now look at the new list of inputs and see if you think it helps." (Wait 30 seconds.) "If we had to choose some more input numbers, what would be good ones to choose?"

Jenny: "How about 6?"

Teacher: "Why?"

Jenny: "It's not there and it looks like it should be."

Teacher: (Puts 6 below 5 in the input column.) "Would anyone care to guess what the answer should be for this input number?" Hold up your hand if you wish to come to the board."

Jon comes to the board and puts 30 below 26.

Teacher: "If I put in 7 as an input number, what answer would there be?"

Maria comes to the board and writes a 34 in the appropriate spot.

Teacher: "Do you remember the silent board game that we did yesterday?" (Copies one

on the board that she had copied down the day before.) "Work on a piece of paper in your groups to see what happens when you put the input numbers in order from smallest to largest."

A five minute group discussion ensues, before it is time for recess.

The following day, the teacher begins.

Teacher: "Today, for our silent board game, I want you to choose the best possible inputs to help guess the answers. Each of your groups has a few minutes to choose the ten inputs you want to make. When you have chosen your ten inputs, one member of your group should write them on the board."

After about five minutes, every group has written its set of inputs on the board.

Teacher: "Now, after you have all looked at all of the sets of inputs, I want you to vote on which one I should fill in."

At this stage, most of the students will recognize that the easiest way to get the answers is to find a pattern and that the easiest way to find a pattern is to organize their guesses in a sequence. Play the game several times focusing on "what is a good sequence of guesses to make?" Once the students seem to have the idea point out explicitly that:

> ANYTIME THAT INFORMATION
> CAN BE ORGANIZED,
> THE PROBLEM WILL BE EASIER
> TO UNDERSTAND.

Once students have the idea of organizing data well in mind, it is time to move on to the second stage of the silent board games. These are played in exactly the same way, but now there are two inputs instead of one. A good way to begin is simply to put the chart at right on the blackboard and ask the class for more inputs, reminding them that they must suggest two numbers, and not just one. (The rule here is that the sum of the three numbers is twenty.)

Unless you have a very unusual class, they will begin by giving all sort of pairs of numbers

Inputs	Answers		Inputs	Answers
			5	
5	9	6	5	
3	4	13	5	
			5	
14	1	5	5	

Figure 6   Figure 7

without any attempt on their part to organize the information. This is normal; old habits die hard. Play a few games with this new format to see if anyone in the class tries to get some organized sequence of guesses. If so, encourage the process. If not, get the process started. One good way to do it is to start a game where you make half of the first five guesses. For example, after putting down the table at right, either let individuals from the class suggest the other inputs or else ask groups to choose what the other inputs should be.

Once they get the idea, point out again how useful it is for them to organize their data in some kind of coherent form to be able to handle it. Talk some about the scientific method and how it depends on holding all variables (inputs) fixed except for one and varying that one in a logical way. If your class did the pendulum activity from Section II, you might also talk about how much easier it was to make sense out of the information when only one variable was changed at a time. If you have not done the pendulum activity, this might be a good time to do it. It will take more than one class session. After one session, pause for some class discussion on what the variables are (length of string, number of washers, force of push, etc.) and then turn the groups loose to finish.

At some point during a class discussion, the new skill of organizing information must be added to the class problem-solving chart, *Understanding the Problem.* The skill of checking your solution with the problem has been used continually. Help your students become aware of this. Suggest that they look for places in which organizing information and checking the solution with the problem helps them in real life as well.

Advanced Headlines and Stories

Introduction

Your students are now familiar with writing stories to match headlines; they know how to operate in groups of four and how to edit. Also they have developed many problem-solving skills and are able to use four techniques: restate the problem, clarify the question, organize the information and check the solution with the problem. Your task is now to prepare them to deal with more complex problems.

There are two parts to this aim. First, they must be able to read and understand more complex stories. Second, they should be psychologically ready for the problem-solving strategies of Section III where they will be investigating problems in complete detail.

There are several ways to make these stories and headlines more complex. The stories may include unnecessary information, assumed information and/or hidden questions. It may be necessary to write two or three headlines to solve the problem. Reevaluating solutions and rethinking methods are necessary at this time. Recognizing, locating, and dealing with each of these will lead to better understanding of these specific problems and all problems in general.

The activities in this chapter are not sequenced as specifically as those in Basic Headlines and Stories. Nevertheless, each technique must be brought to the attention of your students. If the techniques do not occur spontaneously in the stories the students write, then you should introduce a story problem that utilizes them, one at a time. You should introduce each of these techniques through class discussion. Getting through the activities will take four weeks of daily, 15-minute sessions. Three of the techniques developed in this chapter, clarifying the hidden question, getting rid of unneeded information and finding needed and assumed information become additions to the four techniques already listed on the chart.

The fourth technique, evaluating the solution and rethinking methods when necessary, becomes the fifth and final technique listed on the chart. It will be easy to re-order the list near the end of this chapter so that the headings are in logical order.

Understanding the Problem
I.   Restate the problem
II. Clarify the question (includes the hidden question)
III. Organize the information
   1. Get rid of unneeded information
   2. Find needed and assumed information
IV. Check the solution with the problem
V.   Evaluate the solution and rethink the methods when necessary

Suggested Classroom Sequence

On the first day of classes in January begin math class with a story of your own. Tell it aloud and also pass a printed copy to each group of four. It must be a long story, containing many numbers and much unneeded information. This is, of course, on purpose; there is an ulterior motive. Here is a sample story.

#171 Going To The Movies

Jay and Chris went to the show. They left the house at 6:45 P.M. They arrived at 7 P.M. Chris bought their tickets and waited in line for 22 minutes. She got $4.64 in change from her $10 bill. When they finally got inside the theater, Jay stood in line to buy popcorn. He was tenth in line. He bought two tubs of popcorn; each one cost 85¢. Halfway through the movie Jay went back to the snack bar to get two orange drinks and another tub of popcorn. Much to his surprise, one drink cost 75¢. The movie was over at 9:30 and Jay and Chris were hungry again. How much money did Jay spend on popcorn?

Ask your students to review the techniques they have learned for solving problems. They should list the following:[1]

---

[1] If they don't, help them.

Restate the problem.

Clarify the question.

Check the solution with the problem.

Organize the information.
Check the solution with the problem.

The many numbers and the length of the story make it difficult to begin. At this point, it is helpful to ask the class, "Are there any numbers or is there information given that you don't need?" Now the light bulbs begin to come on as hands are raised with affirmative answers. As a class exercise, have the students delete all unneeded information. Then ask each group to solve the problem. This is done quickly and efficiently. This new sub-technique is added to the list.

The students are told to write stories of their own, containing much unneeded information. These take several days to write, edit, revise and share within their groups of four. Fifteen to twenty minutes a day is the most efficient amount of time to be spent doing any of these projects.

One of the interesting things that usually occurs at this stage is that needed information gets left out. The students have great fun putting in unneccessary information and sometimes forget the essential. However, the class problem-solving chart is posted, and someone in the editing group is sure to discover and use it as a guideline. The chart also makes for some fine class discussions.

Another sub-technique that must be developed is looking for assumed information, that is, finding needed information not explicitly stated in the story. Sometimes it is obvious; other times it is obscure. For example:

Johnny decided to read a whole set of encyclopedias. As many of the words were unknown to him, he kept a dictionary close by. This particular set had twenty volumes and many interesting pictures and diagrams. He started reading on January 1, 1980 and finished on December 31, 1981. How many days did it take him to read all of the books?

Most students will recognize that the author of this story assumed that everyone knows a year has 365 days. However, did you remember that this particular year is a leap year? Students need to become aware of and look for implicit information.

During this study of advanced headlines and stories, encourage your students to write problems that require more than one headline. Although this may have occurred earlier, in Basic Headlines, actively seek them out now. (Textbooks refer to these as two- and three-step problems.) As students write, edit and solve more complex problems, their anxiety lessens and their understanding increases. Many students never find out that multiple headline stories are more 'difficult' than single headline stories. They may even do one or two headlines mentally, but write out only one.

As students write more complex stories, something called the hidden question may occur. If it doesn't, the teacher must provide an example.

172 Shelf Paper

As Bill was heading off for the Little League diamond his mother tossed him the tape measure saying, "Don't forget that we are supposed to cover the shelves at the snack booth with paper. Find out how much paper we will need." When Bill finished practice, he measured the shelves. Four of them were 8 feet long and 10 inches wide while the other two were 7 feet long and 11 inches wide. Write down what Bill told his mother.

Sometimes students read through the problem too quickly and skim for the question mark. as in this story and others, there isn't one. Students need to be aware of this and learn how to look for the hidden question.

It must be pointed out that any or all solutions must be tested against the information given in the story. In checking their solutions with the problems, students find out if they understood the meaning of the problem and if they used appropriate problem-solving techniques. There will certainly be times when the solution does not fit the story; when this occurs, the technique called "reevaluate and rethink your methods" comes into play. Although this is listed as a separate strategy, it is an integral part of "Check your solution with the problem."

Rethinking reduces anxiety and encourages open-mindedness. In fact, it produces continual thinking and reevaluation, not stopping at the "answer," which may or may not be correct. In the situations we encounter daily, we often make decisions and come up with solutions, yet seldom have an "answer key." We continually reevaluate, recheck, rethink, in short, do the best we can with the methods, experience and information available. This system works quite well on word problems as well. Another, perhaps unexpected side benefit is that these same techniques can be put to good use when doing computation. Clarify the question, check on the sign of operation; check solution with problem; and ask if your answer matches. Does it make sense? If not, rethink your methods.

Having successfully arrived at this point, the students should now be using certain techniques to solve the problems. Regular practice and teacher-led discussions are necessary for continued success. Here is a sample of the type of problem your students should be able to solve.

173 Spydyk's Birthday

The spaceship, Star Quest, is beginning its sixth month in space with a celebration of Captain Spydyk's birthday at 1800 hours. Alice and Mark were in charge of decorating the main control room for the party as it was the largest room on the ship. To make more room, they were moving out three supply boxes which each weighed 80 pounds on earth.

Alice told Mark she would be there 45 minutes early so they could get everything done on time. Mark wrote the time on a slip of paper which he put into his pocket. Later, he decided to be there 20 minutes before Alice to get the supply boxes moved.

He arrived just when he wanted, moved the boxes and then helped Alice decorate. It was a wonderful party and everyone enjoyed themself thoroughly. But Mark could never remember when he arrived. When *did* Mark arrive?

Solving the Problem

Preliminaries

The work of the two previous sections has laid the foundation for successfully teaching problem-solving. Your students should now be eager to solve problems and be able to strip most problems down to their essentials. Suppose, however, that they still cannot solve a problem which they meet—even after they have completely understood it? Where do they begin? The function of this section is to help them get started solving a problem when they do not know what to do. Several strategies are presented to MAKE PROBLEMS SIMPLER and some guidelines are presented on how to *choose* which of the strategies might help.

> *TO ATTEMPT THIS SECTION WITHOUT THE FRAMEWORK ESTABLISHED BY THE PRECEDING SECTIONS IS TO COURT DISASTER.*

The goal of this section is to guide the students into discovering and then *consciously* using certain strategies to solve problems. In the back of this book the problems are in a form suitable for duplication which you may give to your class. The problems are grouped with the idea that all problems in one group may be solved by the same general strategy.

Students should learn to identify this strategy (probably with some assistance from you) and also, as they learn more strategies, to select the strategy appropriate to the problem.

Five major strategies will be developed.

1. Find and solve a subproblem(s) or hidden problem(s).
2. Use a picture, a map, a diagram, a chart, manipulatives or dramatization.
3. Use easier numbers, estimation, and/or fewer steps.
4. Look for patterns.
5. Work backwards.

While this list is not exhaustive, students who can employ these strategies with reasonable facility will be able to solve most problems they encounter.

To help your students learn these new strategies, present one of the problems from the back of the book to your class every other day or so. Review the techniques which were developed in the preceding section for **understanding the problem.** Then review any strategies for **solving the problem** which have already been practiced. When a new strategy is being developed, it is often helpful to tell your class that they may need to use the new strategy in order to be able to solve the problem efficiently. Introduce the new strategy and then let each group of four go to work—understanding the problem, analyzing the problem and searching for solution strategies.

Next day, have the groups of four share with the whole class the methods they discovered and their approaches to the problem. Allow some class discussion about which groups' strategies are familiar ones and which are new; possibly there will be two or three different descriptions of the new strategy by different groups. Then the new strategy should be added to the class chart on Solving the Problem. (It is probably best when you begin this section to create a new chart 'Solving the Problem' to hang alongside the chart on 'Understanding the Problem'.) If after about

five minutes several groups do not know how to proceed, stop the class and demonstrate the strategy. (You will "hear" the absence of reasoning long before the five minutes are up.)

Each of the problems given in the back of the book takes about two fifteen-minute sessions for completion: one session to solve the problem in the group of four and another session to share methods and process. If the strategy is a new one, a third period may be necessary for discussion. After the students have successfully completed three or four problems using a specific strategy, have them write a story of their own whose solution is most easily found by using the strategy under consideration. These stories may be written by individuals or by groups of four. With this time table, each strategy will take about two weeks for adequate development.

One problem that will probably occur as you proceed is that some group, perhaps several, will refuse to use the strategy which is being introduced. Indeed, as your students develop their abilities, several strategies may be applicable and a group may opt to use one which they are more familiar with. This is a delicate situation. On the one hand, we wish to reinforce the notion that there are many correct ways to solve a problem. On the other, these solutions strategies are specific *skills* which need to be learned. In general, our approach has been to encourage groups to use the particular strategy, but not to demand it.

Many teachers have philosophical problems with the schedule outlined above. A common reaction is to feel that far too much time is being spent on just a few problems. Surely if doing five problems is good, doing ten problems is better. Not necessarily. In fact, as we have learned from painful experience, if the ten problems are considered in the same time as the five, students will probably learn *less*. Why? Because, students who only grasp dimly what is going on in the solution and cannot correlate solving this problem to others of a similar type, will quickly lose the little knowledge they have gained. One of these will be the student who swears to you two weeks later that he has never, ever, in his whole life, seen this particular problem before and will be most surprised when you hand him his own paper with the solution to the problem on it.

Suppose on the other hand that the problem is carefully considered from several points of view, that alternate solutions are given which illuminate various aspects of the problem and that the students have the time to tie this solution into past knowledge. Now the ideas have a chance to stick, a chance to be assimilated and integrated, and something real is added to the students' intellectual structure.

This idea parallels the the old Chinese saying:

> If you give a man a fish, you feed him for a day;
> If you teach him to fish, you feed him for a lifetime.

In our case the *process* of finding a solution is far more important than the answer itself. No individual problem has an answer important enough to be worth just telling. But the reasoning process behind the answer is worth lots of time and energy to get across. Your students will probably (even this late in the year) have difficulty with the idea that you care about the process and not just the answer. After all, this flies in the face of everything they have learned during their school careers. You will have to convince them that you mean it. You will probably have to change your grading policies to reflect this viewpoint by having the answer count for little and the explanation count for a great deal. This is another reason for giving a few problems, for neither you nor your students are going to be willing to do all of the work that is required to exemplify this point of view for more than a few problems per week.

To facilitate your students' efforts throughout all of this learning of strategies, keep the two charts on MAKE IT SIMPLER posted in a prominent place. If all goes well, at the end of the year, your charts should look like those below.

Make It Simpler

Understanding the Problem

I.   Restate the Problem.

II. Clarify the question (includes the hidden question).

III. Organize the information
   1. Get rid of unneeded information;
   2. Find needed and assumed information.

IV. Check your solution with the problem.

V.   Evaluate the solution and rethink your methods when necessary.

Solving the Problem

I.   Find and solve a subproblem(s) or hidden problem(s).

II. Use a picture, a diagram, manipulatives, or dramatization.

III. Use either numbers, estimation, and/or fewer steps.

IV. Find and use patterns.

V.   Work backwards.

How the Problems Are Numbers. Almost all of the problems given for use with the five basic strategies are presented in three different forms at three levels of difficulty. The numbers given to the problems reflect the relationships. For example, problems 115, 215 and 315 are essentially the same problem (when seen from a sufficiently advanced viewpoint). Problems with numbers in the 100s are easy, 200s are average and 300s are advanced.

While the hundreds' digit reflects the difficulty of the problem, the tens' digit tells the *primary* strategy useful in solving the problem. Some children may solve the problem in a very different way and that is fine. However, most groups will probably use the stated technique. The correspondences between the tens' digit and solution strategy are:

1. Find & solve a subproblem/hidden problem;
2. Use a picture, diagram, manipulatives or dramatize;
3. Use easier numbers, estimation, and/or fewer steps;
4. Find and use patterns;
5. Work backwards;
6. Problems with multiple solution;
7.
8. } Mind-expanding problems.

Thus problem 243 is of average difficulty and will probably be solved by finding a pattern. The ones' digit represents the suggested sequence.

C H A P T E R   8

Strategy 1: Find and Solve the Subproblem (Hidden Problem)

Preliminaries

This first technique is almost universally applicable and is a natural outgrowth of the Advanced Headlines section. There your students were encouraged to write stories with hidden questions and multiple headlines. In essence, whenever a story requires several headlines, each of these headlines solves a "subproblem" or part of the stated problem. In general, students tend to be more comfortable with the term "hidden problem" with its connotation that the writer has cleverly hidden an extra problem that needs solving in the middle. We tend to use the terms interchangeably with students although "subproblem" is more proper.

Your students have found and solved hidden problems even before these. When they played "Poison," the optimal strategy involved solving a succession of hidden problems.

> Problem: How can we get exactly one block left?
> Answer: By getting to the stage with four blocks left.

Hidden Problem: How can we get exactly four blocks left?
Answer: By getting to the stage with seven blocks left.

When they played "Color Square," generally they began by solving the hidden problem of finding the color of the square in one corner. Once that hidden problem was solved, they were then able to solve the hidden problem of the color in the next square, etc.

So solving subproblems is something your students have done a lot of without ever recognizing that they have done so. Point out to them these many instances where they have solved subproblems. Pick up some group's story with a multiple headline, and ask them to identify the subproblem which gave rise to each of the various headlines. When you have finished all of this preparation, they should know what a subproblem is.

The next step is to get your students actively to look for the subproblems. Then, even if they are unable to solve the original question, they can MAKE THE PROBLEM SIMPLER by solving each hidden problem in turn. Every subproblem should be reviewed as a problem in its own right worthy of time and effort. Some of them will be trivial, but usually solving one of the subproblems is the key to solving the original questions. Learning to identify that central hidden problem is a crucial step to becoming a good problem-solver.

You will find that your students are often reluctant to solve only part of the original question; as yet, they usually do not see the value of this step. So it is up to you as their teacher to convince them of the importance of this process. You may refer to the "Color Square Game" as an example of how doing part of a problem helped to do the rest. Your class did not need to know the entire pattern in order to decide what color a single square was. Nor do they need to be able to solve an entire problem in order to solve a part of it. Moreover, once that part is solved, it should be SIMPLER to solve the rest of it.

Because of the importance of this strategy and because this is the first problem-solving strategy which they will learn, you probably should help your students with the first *three* problems they do. (Recall that we suggest a different approach in general.)

Plan on spending fifteen to twenty minutes doing the first problem together from beginning to end as a whole class. As you go through this first problem with your students be sure to elicit the subproblems from them, writing each of them on the board. As they solve each of these questions, they should begin to see more clearly how useful this technique is.

Do the second problem almost to the end, writing down the subproblems and then turning the groups loose to solve them. For the third problem, go as far as writing down one hidden problem and leaving the discovery of the others and the solutions of all to the groups.

#211 The Hrunkla Village

Captain Spydyk stepped out of his space ship, the Star Quest, on the surface of the alien planet. Hundreds of small creatures resembling furry snakes surrounded his force field while his language-translating computer listened to the strange whistles coming from them. Within 10 minutes, the creatures had arranged themselves into a large rectangle, 24 rows deep with 25 creatures in a row, and were speaking in unison. At this point, the computer began to be able to translate the key features of what was being said.

It seemed that the Star Quest had landed on top of several of the homes of this village, completely ruining them and the creatures of the village made up the rectangle which was talking to the space ships. This village consisted of 35 living groups each with 17 Hrunkla plus some Mankla (high rulers) who lived alone. All of the Hrunkla homes were safe, but each of the Mankla, who lived alone, needed another home to be built. It should not take long, but the Star Quest would have to move.

How many homes were needed for all the Mankla of this village?

Sample Dialogue

Teacher: "Will someone please restate the problem"

Greg: "There's a village and it needs more homes or something."

Teacher: "Will someone please restate the questions?"

Debbie: "How many of these Mankla need to have new houses."

Teacher: "What steps have we learned to help us understand the problem?"

Atsuko: (Points to the list.) "But I don't think any of these are very helpful. I understand the problem. I just don't know where to start.'

Kurt: "I don't get it either."

Dan: "I think we should subtract something from something else."

Allison: "If we knew how many creatures there were in all, that would help."

Teacher: "Allison's idea is part of the way to solve this problem. If, somehow, we knew how many creatures there were in all, it would help us get the answer. Finding the number of creatures is a hidden problem or subproblem which we have to solve in order to answer the question asked.

This idea is not really new. Remember last week when we had stories with multiple head-lines—each one of those headlines solved part of the problem. And you had hidden problems when we played "Color Square"; each square was a separate subproblem to be solved. So you have looked for, and solved, hidden problems lots of times." (The teacher expanded on this theme.)

"Now I want you to find more hidden prob-lems in your group."

**Allow two to three minutes of discussion.

Teacher: "Have any groups found some other hidden problem?"

Oran: "Yes. Another one is to find out just about how many Hrunkla."

Teacher: "Any more? Or is this all?"

Jocelyn: "I think that's all."

Teacher: "Now that everyone has a place to start, I'm going to give you some time to work on the solution in your teams."

**After five minutes or so—

Teacher: "I would like to hear from as many teams as possible. Tell us about a hidden problem you solved; what was the first one?"

Team 3: "We found out that there were 595 Hrunkla."

Team 7: "We found out that there were 600 creatures in all."

Team 8: "The rest was easy; you just subtract 595 from 600."

Team: "How did you get 600 in all? That was the part we couldn't do."

Teacher: "So that was a subproblem for you: find how many creatures in all. Now that you see it as a problem, can you solve it?"

Team 1: "We got 800 in all. Who's right?"

Teacher: "You know me better than that. I expect each team to prove their answer is correct."

Team 1: "Never mind. We multiplied wrong."

Team 5: "O.K. Now we get 600. But that was hard. You had to do a lot of different things."

Teacher: "Yes. It was a hard problem. But each part was simple. And you did solve it."

Although this seems to be somewhat redundant, it is a close replica of a discussion by a real class of fifth graders. What is obvious to an adult does not appear to be obvious to students, and we feel this interchange has great value on the overall goal of developing the students' problem-solving abilities. We include it for this reason.

Problem 311 is similar in nature. The dialogue will be somewhat the same, the math a little more difficult. Choose the problem that best fits your students.

Here is the next problem, which contains sub-problems. It is easier than the first.

#212 Settling Up the Bills

Four students decide to throw a party to cel-ebrate the end of school and share the expenses equally. Amy buys a cake for $8, Randy buys $5 worth of ice cream, Sarah spends $2 on green and red crepe paper to hang, and Dave gets $3 worth of soft drinks. In addition, he pays the $6 to rent a giant popcorn popper. To be fair, who owes money to whom?

Comments on #212

The central subproblem here is finding out how much money was spent in all. From there, finding each person's share is easy, and then it is easy to determine who owes whom what.

Some students, however, don't even try to make it come out exactly. Views range from "We can make it up next time" to "If Amy and Bill each give Dave two dollars, then it's all fair. Students at a slightly more sophisticated level can calculate exactly what each per-son's share should be, but get lost in trying to go from that fact to answering the questions.

The comments on the two problems above give some indication about how difficult it is to describe all the possible approaches students use to solve problems. So we will not try. On the back of each problem in Section IV and V are comments giving typical approaches to the problem together with some hints. We could easily list three times as many solution approaches, but it does not seem worthwhile; your class will probably have a group who does it differently from any listed anyway. We trust that these comments give some idea of our expectations and hope that they match yours.

Strategy 2: Use A Picture, Diagram, Manipulatives, Or Dramatization

Preliminaries

The problems in this section are designed to be simplified by drawing a diagram or a picture or by using objects or drama to help solve them. Frequently, just shifting the setting of a problem from words to a picture is enough to help students solve the problem. Recall for them that when they played "Poison" sometimes it was easier to figure out the strategy when they played with blocks or squares instead of just numbers, even though the three formulations of the problem were all equivalent. Drawing a diagram is also a useful technique for understanding the problem. If the language is confusing or if it is clear that geometry is involved, a diagram is frequently helpful. From this point on, you should *actively encourage* your students to draw a diagram or use objects or drama *whenever* they feel it would help them.

Often a *subproblem* can be easily solved with a diagram or manipulating objects. After arriving at this solution, a headline approach may be used to solve the original question. Helping students make decisions on the strategies needed is a central part of this entire section.

Occasionally you will run into students who are so bright that they feel it is below their dignity to draw a picture. Make them do so. If Einstein felt that diagrams were useful to him, your students should not feel they are beneath them. A more usual kind of reluctance comes from the students who simply does not want to spend one minute drawing a picture. For these students you might hold out the results of a study some years ago which showed that students who draw pictures solve problems better and *faster* than those who do not.

In any event, all of your students should be comfortable with the idea that *whenever pos-*

sible a diagram should be drawn. It not only helps them understand and solve the problem, it may also help them check their solution.

Activity

The dialogue problem for this section is suitable for all three levels. Fourth graders can do it after thought, while a few eighth graders solve it immediately.

#221 The Commemorative Scroll

In order to demonstrate his peaceful intentions, Captain Spydyk instructed Lieutenant Schwartz, who was an accomplished artist, to prepare a poster or scroll commemorating their arrival on Smygiaa. The message was to be put on a piece of white posterboard 38 inches high and 48 inches wide. It was decided that each letter should be two inches high and that a one inch space should be left between the bottom of one line of words and the top of the next line. In addition, there should be three inch borders at both the top and the bottom of the posterboard. How many lines of words could Lieutenant Schwartz fit onto the poster?

Sample Dialogue

Teacher: "Will someone please restate the problem?"

Kuo-Wei: "The man has to fit the words on a poster."

Teacher: "What is the question?"

Yolanda: "How many lines does he have to write on?"

Teacher: "What strategies have we practiced so far?"

Craig: "Getting rid of useless information."

Michelle: "Looking for subproblems."

Teacher: "Will any of these help?"

Pat: "It's too confusing."

Teacher: "What might we do?"

Rachel: "Organize the information."

Teacher: "How?"

Pat: "It's too confusing."

Jim: "We could draw a picture."

Jason: "There's too many inches."

Teacher: "I have some graph paper—if anyone would like to use it, you may."

Although this sounds rather simplistic, it is almost verbatim from a fifth grade class. Many of my students do not seem able to look around for props or materials that might prove useful in solving a problem. The suggestion of graph paper (¼ in.) was enthusiastically received and all groups were quickly at work, solving the paper.

(Note: Even with graph paper, several teams had incorrect solutions the first time. I needed to remind them to check their solution against the stated problem.)

Strategy 3: Use Easier Numbers, Estimation, and/or Fewer Steps

Preliminaries

This next technique is the simplest of all to use. It is easy for students to understand and they can apply it immediately once they hear the idea. In addition, it frequently leads directly to a solution. The only difficulty which you will have in getting students to use it is that the idea seems too simple to be worth doing. At the beginning they may object that they are not solving the "right" problem. (On the other hand, if they can solve the "right" problem, they do not need to use any of the strategies.) So remind them constantly to use this technique, either alone or in conjunction with some of the others.

The idea is simplicity itself: replace all of the hard numbers in the problem with easier ones. In the sample problem for which the dialogue is written, the question is seen to be elementary once the fearsome large numbers are out of the way. Then—once the students know the *process* for solving the problem with simpler numbers—they can go back to solving the original problems.

An example may help. Tom's nine-year-old daughter was given the following problem: How many 1' x 1' square tiles would it take to cover a floor which is 10' by 5'? She replied, "I don't know." Then she was asked how many tiles it would take to cover a floor which was 2' x 3'. Her answer was "50." Here she solved the easier problem, understood the process and then used the same process for the more difficult problem. This is the sort of thinking that students should eventually be able to do for themselves.

The numbers may not need to be smaller to be easier. Many students find it more natural simply to approximate the numbers in the problem and thus *estimate* the answer. Such an approach is the best possible; not only do these students discover the proper procedure by this method, but they also know about what the answer should be.

Students who would prefer to use smaller numbers will still get some insight into the solution techniques if they choose their numbers carefully. In general, the number 1 is not very good because it may mask the difference between dividing and multiplying.

Other one-digit numbers are generally good; so are multiples of 10 or 100. Students should be encouraged to round fractions to whole numbers; they should use anything which eases the computational burden.

Some children may need to solve a sequence of several simpler problems until they understand exactly what the underlying process is. In a sense, these students wind up observing the workings of their own minds as they see how they themselves solve the simpler problems. Such an activity is not only an extremely useful learning experience, but also a great ego booster.

The same technique for making numbers easier can also be used to good effect in problems involving several steps (for example, 234 or 334). In this case it is easiest to solve the problem with only one step. Then solve the problem for two steps, etc. and so on.

Ready? Here is the first problem.

Activity

#231 Captain Spydyk and the Jewels

Captain Spydyk looked out at the new part of the alien planet on which his space ship had landed the night before. A huge line of the native creatures (Hrunkla), were waiting outside the force field protecting his ship. By this time his language-translating computer was able to make some sense of the "speech" and was translating parts of it.

In gratitude for Captain Spydyk relocating the space ship, all 27,804 creatures in the area had contributed to gifts for the crew. As was their custom, each creature had contributed a tiny diamond-like jewel. Pyramids were made out of clusters of 84 of these jewels and lucky individuals in red robes were waiting patiently to give these to Captain Spydyk.

Captain Spydyk worried about the weight involved in accepting so many gifts, even though the jewels looked valuable. But it turned out that a pyramid only weighed 23 grams,

so he realized that all of them together would weight only what?

Comments

This problem is difficult enough to be used at the more advanced level also (as 331). However, the arithmetic skills of some younger students are not adequate to handle the necessary multiplication. If this is true of your students, use the following in #131: 20,000 creatures, 50 jewels per pyramid and 20 grams.

Sample Dialogue

Teacher: "Now that you have read the story, we will get ready to solve it. First, who can restate the problem?"

Felicia: "They are trying to find out how much all of the jewels weigh together."

Sergio: "No. I think they are trying to find out how much the pyramids weigh."

Teacher: "Which? We had better decide."

Jenny: ""Pyramids. That other stuff about jewels is just to trick us."

Stacey: "No. It's important. But I don't see how."

Teacher: "All right. Is there any unnecessary information."

John: "Yes. That stuff about what the funny things look like, and the speech, and a lot of other things."

Teacher: "All right. Work in your groups to get rid of any more useless information and tell me when you have the problem solved."

Mary: "Can we use calculators? Those are big numbers."

Teacher: "When your team can tell me what you want to do, then you may use calculators."

***Long pause—several groups may be discussing how to proceed. One minute is a long time.

Teacher: "Stop, everyone. If you haven't solved the problem yet, I have a strategy which is frequently useful. All you do is solve the same problem with simpler numbers. Let's assume that there are only 10 Hrunkla, that each pyramid weighs 6 grams and that there are only 2 jewels per pyramid. Try solving this easier problem with your group."

****Put new information on the chalk board or overhead. Wait two to three minutes.

Teacher: "How did you do?"

Team 4: "That's easy. The answer is 30 grams."

Teacher: "For our purposes, the answer is not important. What did you *do*? We have a different problem to solve."

Student in Team 6: "We found out that there are 5 pyramids and multiplied by 6 to get 30."

Teacher: "How did you get 5 pyramids?"

Kurt: "We divided 10 by 2."

Teacher: "Did any team use another method?"

Team 3: "We divided 6 by 2 to get 3 grams per jewel then we multiplied 3 by 10 to get the total number of grams."

Team 5: "We still can't do it."

Teacher: "Let's make the problem even simpler then. Suppose that there are 4 Hrunkla, 3 pyramids contain 2 jewels and weigh 1 gram each. Try this version and then try the other one I gave you."

Team 7: "We're almost done with the *real* problem but need a calculator to check our multiplication."

Teacher: "O.K. I think most of you can see how making the numbers easier helps to solve the original problem. You can get calculators and proceed. I will be asking each group *how* they got the answer not *what* the answer is."

***Have a five minute work period followed by oral team reports as promised. Ask students to restate strategy and add to your wall chart on solution strategies.

Strategy 4: Find and Use Pattern

Preliminaries

This problem-solving technique exemplifies many of the ideas which we have already developed. Recall with your students how much easier it was to guess the "rules" in the silent board games when their numbers were arranged in a pattern. Looked at another way, patterns are a continuation of the idea of making numbers smaller—except that here you need a *succession* of smaller numbers.

Your students should already have the idea; they merely may need someone to point out that this practice is a *bona fide* problem-solving technique. Here, if you cannot solve a problem with the numbers that are specified, *make the problem simpler* by using a sequence of smaller numbers until you see a pattern emerge. Then use that pattern to *guess* the answer to the original problem.

Note that this technique has a very different flavor from the other strategies. With each of the first three strategies, the students were led to discover a technique which applied directly to the solution of the problem. Here,

on the other hand, the students will generate a (possibly wrong) guess about what happens in complex cases based on their experience in simple cases.

Several trial numbers are needed in order to detect a pattern. Students who try two or three numbers and claim they have found the pattern *must* be encouraged to go back and try some more numbers to check and be sure. It is far too easy to guess the wrong pattern on the basis of limited information. A reasonable minimum number of guesses would be five. The four students in a group may very well choose to have different individuals make the computations for different cases and go from there. In such circumstances it is always well to have two students do each computation so that one wrong number does not completely destroy their chances of finding a pattern.

What types of problems are amenable to solution by the use of patterns? Virtually any problem is. Subproblems may be solved by using patterns as well. In general, if you look at a problem and have no idea what to do, no subproblems are apparent and one set of smaller numbers still leaves you in the dark, try looking for a pattern. It may jump out at you after only a few guesses. If not, keep at it until you have generated at least ten numbers in a sequence.

You might want to play a silent board game immediately before introducing patterns to be sure the idea is fresh in your students' minds.

Activity

Our first problem is suitable for all levels of students and is rich in a variety of patterns. Enjoy it.

#241 Handshake

If *everyone* in our room shook hands with everyone else, how many handshakes would there be?

Sample Dialogue

Teacher: "Will someone please restate the problem?"

Yolanda: "Everybody shakes hands—how many are there?"

Teacher: "Will someone please restate the question?"

Rachel: "How many handshakes will there be?"

Teacher: "What strategies have we studied so far?"

Michael: "Get rid of useless information."

Teacher: "Do we have any?"

Greg: "No."

Teacher: "Any other strategies?"

Felicia: "Find a subproblem."

Teacher: "Do you see one?"

Kuo-Wei: "No."

Student: "Maybe we could make the numbers smaller."

Teacher: "Will that help?"

Mary: "I think so."

Teacher: "OK, let's do it."

Alison: "We don't have any . . . Yes we do!

. . . How do we count handshakes and people?"

Teacher: "How many people are here today?"

Craig: "31"

Teacher: "Do you want to count me too? Would anyone care to make a prediction before we continue?"

Very few students will; however, you may want to record all guesses that are made with no comment other than a thank you. You might also encourage all students to make a private prediction.

Teacher: "How much smaller can we make the number of people?"

John: "10"

Debbie: "5"

Michelle: "2"

Jim: "1"

Teacher: "OK. Let's start with one. The very beginning. How many handshakes if there were only one person here?"

Oram: "None."

Teacher: "Two people?"

Jocelyn: "One—Two"

Teacher: "OK. We need to know how to count handshakes. In this problem, when two people shake hands, it counts as one handshake. I'm going to set up a chart like this: Which group can show me the number of handshakes for three people?" (All groups should be actively involved in handshaking at this point.)

Stacey: "Three." (I don't call on any group until I see most groups finished and ready to answer.)

Teacher: "Please demonstrate for us."

Students: (Demonstration—class counts)

Teacher: "Now we need to find a way to figure out how many handshakes if *all* of us shake hands. We don't take the time to have everyone shake hands, let's look for a *pattern*. Finding patterns is another way to solve problems."

"Let's make a table like this:"

People	Handshake
1	0
2	1
3	3
4	6
5	.
6	.

"I'm going to give you time now to work on this problem. When needed, you may join up with other teams in order to count handshakes."

*If a group should finish very quickly, ask them questions such as: How many handshakes if three are absent? In both sixth grades together?

*During class discussion of this problem, the new strategy should appear. Bring it out by asking: "What patterns did your group discover while working on this problem?" (Note: This does not ask anything about the *answer*.)

Patterns

Students: "The numbers go up in order."

People	Handshakes	
1	1	0 > +1
2	2	1 > +2
3	3	3 > +3
4	4	6 > +4
5	5	10
6	6	
7		

Michael: "Add across to get the next one."

People	Handshakes
1	0 (1+0=
2	1 (2+1=
3	3 (3+3=
4	6 (4+6=
5	10 (5+10=
6	15 (6+15=

Jenny: "Odd numbers are timeses."

People	Handshakes
1	1 x 0 = 0
3	3 x 1 = 3
5	5 x 2 = 10
7	7 x 3 = 21
	9 x 4 = 36

Teacher: "I see an adding pattern on the left side." (Sometimes this will help a student see it—if not, continue.)

People		Handshakes
	1	0
0+1)	2	1
1+2)	3	3
1+2+3)	4	6
1+2+3+4)	5	10
1 + 2 + 3 + 4 +5)	6	15

"Can any of these patterns be used to find the answer to our problem.?"

Dan: "Yes, the . . . "

Debbie: "Yes, the . . ."

Teacher: "Any of these patterns may be used, some will just take more time than others."

Sergio: "I know a rule: You take the number of people x the number −1 because everyone shakes hands with everyone except themselves. Then you divide by 2 because you need two people to count one handshake.

Encourage continued discussion of pattern and other strategies and processes, asking each group for a "how did you solve it" explanation rather than "what answer did you get."

Strategy 5: Working Backwards

Preliminaries

Our final strategy is the most specialized of them all. It is useful with fewer problems than any of the others, but it still seems worthwhile to point out specifically. Some of your brighter students may recognize this strategy as a special case of using fewer steps or of solving subproblems. Most of your students will not, however, and they will need to have the concept spelled out for them.

The idea is this: certain problems with several steps resulting in a known value can most easily be solved by working backwards one step at a time. In this instance, our sample problem, "The Golden Apples," is a perfect example of the type of problem for which the strategy is useful. The problem is easy if the prince only meets one troll. Then it is easy to solve the problem with two trolls by using the answer obtained for one troll. The students can keep working backwards until they have the answer to the original problem.

Again, the idea is to simplify the problem somehow—in this case by considering one troll at a time. We work backwards from the end because the end is where the concrete information is.

Activity

Enjoy this problem. Students of all ages seem to find it appealing.

#251 Golden Apples

A prince picked a basket full of golden apples in the enchanted orchard. On his way home, he was stopped by a troll who guarded the orchard. The troll demanded payment of one half of the apples plus two more. The prince gave him the apples and set off again. A little further on he was stopped by a second troll guard. This troll demanded payment of one half the apples plus two more. The prince paid him and set off again. Just before leaving the enchanted orchard, a third troll stopped him and demanded one half of his apples plus two more. The prince paid him and went sadly home. He had only two golden apples left. How many apples had he picked?

Ask the class to read the problem silently.

Sample Dialogue

Teacher: "Would someone please restate the problem."

Kurt: "The trolls took all the Golden Apples except two."

Teacher: "What is the question?"

Jason: "How many apples did the prince pick?"

(If the students cannot restate the problem and/or ask the question, go back to "Understanding the Problem"—Chapter 5.)

Teacher: "What strategies have we practiced so far?"

Oran: "Make the numbers small."

Rachel: "Act it out."

Yolanda: "Draw a picture."

Dan: "Look for patterns."

Pat: "Identify subproblems."

Teacher: "Which of these might be useful?"

Long pause—(at least 10 seconds—count if necessary, but wait!)

Teacher: "Which of these is not useful?"

Mary: "You only gave us one number, 2. That should be small enough."

Jim: "Drawing a picture sounds dumb."

Felicia: "I think there is a pattern, but I don't know how to say it."

Teacher: "Can anyone help Amanda?"

Allison: "Each troll takes half plus two more—there are three trolls."

Greg: "You can only have two halves!"

Teacher: "No one has mentioned our overall strategy, 'Making It Simpler.' What changes could we make to make the problem simpler?"

Kuo-Wei: "One troll."

Teacher: "Can you restate the problem using one troll?"

John: "A prince had some apples. A troll made him pay half of the apples plus two more."

Teacher: "What is the question?"

Stacey: "How many apples did the prince have?"

Teacher: "Now I'm confused—there must be some information left out."

Atsuko: "The prince has two apples left."

Teacher: "Thank you—each team try solving this one troll problem."

five-minute interlude (noisy)

Teacher: "OK, everyone quiet; how did you solve this problem?"

Team 5: "The prince had two apples. Just before that he gave the troll two apples, so he had four. Just before that he had given the troll half, so four is half of eight so the prince had eight apples."

Teacher: "Who else solved the problem? (All teams did.) Did anyone use a different method?"

Team 3: "We tried starting with six apples. It didn't work, so we tried eight from what we learned, and it worked."

Team 7: "We are almost done with the real problem . . . you take —"

Teacher: "Hold it. We need to check, evaluate, the solution. How do we do that?"

Jenny: "Tell the story over, using the number. A prince had eight apples. He paid the troll half (four) plus two more (= six). He had two left. It works!"

Teacher: "Any questions? (wait at least six seconds) OK, now. You have a new strategy: working backwards. Each team may now try to solve the original problem."

(Any group not solving this in about 10 minutes will come "unglued" so you may stop all group discussions and conduct a class discussion.)

Questions for Discussion

1. What strategies did you use? (Make a chalkboard list. "Working backwards" should come up; if not, bring it up.)
2. How did you evaluate your solution? (Make a chalkboard list.)

3. Do you believe your solution is the only correct one? Why or why not? (Complete a chalkboard tally.)

If all groups are sure their solution is correct, you may ask for each solution in written form. The class may then begin another activity. You should check each solution in private, arranging time to meet with any group whose solution is incorrect.

If some groups are not sure, discuss why they aren't as a class and offer suggestions. A great deal of learning seems to take place at this point, often within all groups, as the verball explanations of the strategy clarify the thought process for the speaker.

C H A P T E R   13

Selecting the Right Strategy

Preliminaries

About the middle of April, it is *time for you to decide* how to spend the last six weeks of the year. The choice is whether or not to move beyond the skill of employing strategies to the most sophisticated skill level: *selecting* the right strategy. Generally speaking, if your class has been absorbing the various strategies easily and are confident of their abilities, move on to this new level. On the other hand, if some students are not sure how to find hidden problems or, if they have difficulties detecting patterns, do not try to move on to this more sophisticated level, for it would probably be counterproductive.

Choice 1. You decide not to introduce the final skill.

Because the needs of your class are best served by not teaching the techniques of this last section, we suggest that you proceed as follows. Use additional problems from the back of the book similar to those you have already discussed with an emphasis on the areas

where your students are weak. Your most important jobs as a teacher are to select the problems carefully and to moderate the classroom discussions wisely. For an occasional change of pace, have your students make up their own problems or play a logic game. Keep reinforcing the strategies and strive to end the year on a successful note.

Choice 2. You decide to teach the art of selecting the right strategy.

This topic will probably require six weeks. Do not even consider trying the ideas in less than four weeks. Why are we so insistent on this point? Because starting this topic will unsettle your students somewhat in their approach to problems, and they must have time to establish a new internally consistent way of treating problems before the year ends. If they begin the summer in a new state of disequilibrium, much good may be undone.

Your allocation of class time will not change. Each problem will usually require one session for group work and solution and one session for class discussion. Occasionally the groups may request a third work day to reevaluate their solutions and rethink their methods. Go ahead. This is certainly a productive use of class time.

What will change, however, is the focus of the discussions. The new goal is to encourage the students to review automatically all of the strategies they know and then to focus on those which are most likely to be successful for a given problem. Eventually, this review should become a well-ingrained habit, done unconsciously.

The first step, of course, is for the process to be a *conscious* one, with group work and class discussion focused on decision-making issues: "What strategy looks best for this problem?"; "What was it about this problem that made you think this strategy might work?"; "Did you think another strategy might be useful? Why or why not?".

As you use this approach, keep in mind that there is no mechanical procedure known for selecting the most appropriate strategy, and it is a huge disservice to the students to suggest otherwise. Solving problems is an art, not an exact science. Moreover, the strategy which appears absolutely obvious to you may not be at all obvious to one of your students— or to a fellow teacher. Different people organize information differently and because of this will approach problems differently. So help your students develop their *own* ways of attacking problems, based on the models presented in class. These ways are the ones which will remain with them forever.

Activities

In order for your students to get the greatest benefit from this final topic, you will need constantly to reinforce the notion of first looking for a good approach, and then selecting the specific way to use the approach. That is, your students' first step in solving a problem should be to identify the particular general strategy which they intend to use, then they should concentrate on trying to make the strategy work. In order for this idea to be effective and to help your students resist the temptation to

search randomly, require them FIRST to choose the general strategy and to WRITE IT DOWN. At that point, they should proceed to work with that strategy to find a solution to the original problem. If they change their minds about their first general choice, have then *write down* their new general strategy and begin again. The underlying message in all of this should be that it is SIMPLER to first identify the general strategy than to hunt for the particular solution. Once your students decide on the strategy to use, then they will more easily see the appropriate version of the strategy under consideration.

Thus, the initial focus of discussions within the groups should be about the general type of problem. What is it about the problem which suggests a certain strategy? Although your students won't think in these terms, they are really looking for underlying mathematical structures and will start saying "This problem is a lot like . . . ." As teacher, you may encourage this process by giving problems similar to those which have already been discussed in class. (Recall that if you have discussed problem 134 then problem 234 is quite analogous.)

You should review the general strategies frequently and try to apply them in as many nonmathematical situations as possible. If you are planning a class party, many subproblems emerge which need to be solved. Designing props for a play illustrates the importance of diagrams. Trying to read a complicated set of directions is a good time to review strategies for understanding the problem. You will be surprised at how many places there are to use these ideas.

Still another way of developing "strategy sense" is to give your students a mixed collection of problems and ask them quickly to decide which strategy should be used. An easy problem can simply be read to the class aloud. Then after giving them time to digest the question, call for a show of hands on the appropriate strategy. Discussion following the vote is often useful. A similar exercise may be done in the groups by passing out problems on cards.

In all of these activities, be sure to give as wide a variety of problems as possible and continually support your students by encouraging them, asking questions, focusing discussion but NEVER telling answers.

Homework problems assigned during this period seem to be most useful if they require naming the primary general strategy used in addition to the solution itself. Don't be too surprised to see the same solution characterized as an example of two or even three general strategies; experts often disagree among themselves as to what primary strategy is exemplified by a particular solution.

For a change of pace, have students write a story illustrating a particular problem-solving strategy or just an interesting problem. You might also assign two or three problems to be debated in one period, but only to the extent that the groups would try to identify the probable strategy which would be useful in solving the problem.

Our list of strategies is not exhaustive and is not meant to be. Other books mention different strategies and you may wish to discuss some of them with your class. Probably the most common technique which we did not cover in detail is *Guess and Check*. The idea is inherent in the name. Try a number. If that number does not lead to the right answer, try a better number. The skill lies in devising a well-organized approach to selecting better numbers.

In addition to all of the strategies you find in books, imaginative students will discover some of their own. Such a discovery can stimulate a class discussion on whether or not their strategy really is new. It should also stimulate others to look for new strategies; students ever on the lookout for more strategies are bound to be better problem solvers.

The following dialogue represents a fairly typical example of what happens when you first begin to look at the idea of selecting the right strategy. The problem seems to be a popular one with lots of scope for different approaches.

#371 At The Zoo

The keeper of the bird cages at the zoo discovered that two crested cockatoos would eat two pounds of bird seed every two weeks; that three Peruvian parrots would eat three pounds of bird seed every three weeks; and that four Mozambique macaws would eat four pounds of bird seed every four weeks.

How many pounds of bird seed will 12 crested cockatoos, 12 Peruvian parrots and 12 Mozambique macaws eat in 12 weeks?

Sample Dialogue

Teacher: "Today we begin the last part of our work in problem-solving. You have learned how to read problems carefully and know five strategies for MAKING PROBLEMS SIMPLER, which are listed on our bulletin board. Now the question is, can you decide which strategy to use? Here is a brand new problem for you to discuss in your groups. I am not so interested in hearing a solution today as I am in finding out which strategy you would use. Remember—first decide on your strategy."

*Students work in groups for about five minutes. Teacher writes abbreviations of strategies on the board.

Teacher: "All right. I'm sure you are not done yet, but you should have decided which strategy you would use. But first, what is the question?"

Kurt: "How many pounds of bird seed get eaten by all of those birds in twelve weeks?"

Teacher: "Which strategies look promising—or don't look promising?"

Michelle: "I don't think we can work backwards. There isn't any backwards."

Teacher: "Do the rest of you agree? Or does someone want to try working backwards?" (After a pause, she crosses "working backwards" off the board.) "Who is next."

Atsuko: "I think it is a pattern problem."

Teacher: "All right. We might try patterns."

(Puts a check by "patterns" on the list.) "Any other ideas?"

Jocelyn: "I thought we could just use easier numbers. You know, like maybe only one of each kind of bird."

Teacher: "There is another idea." (Puts a check by "use easier numbers.")

Craig: "It is all about subproblems. One subproblem is to find out how much seed the parots eat. And like that."

Teacher: (Puts a check by "Subproblem.") "Well, we have thought about four of our five strategies. Does anyone have any feelings about using diagrams or manipulatives?"

Sergio: "How can you draw a picture? That's impossible."

Jenny: "I guess we could get some blocks for bird seed and do the problem that way, but it seems like a lot of counting."

Teacher: "It sounds, then, as though there are lots of different ways to solve this problem. Now CHOOSE one of the strategies we have talked about and work in your groups on solving the problem USING THAT STRATEGY. Tomorrow, we will talk about the different solutions and which general strategy seemed to work best on this problem."

Many additional problems are listed under the heading of Mind-Expanding Problems. After the index listing of each of these problems is a list of the strategies which are most likely to be used to solve the problem. Thus, even if your students have not completed their study of all five strategies, you will still be able to select quickly those problems which your class can use. We hope you will find some new problems here that you have never seen before, so that you can share with your students the challenge and the joy of solving problems. Good luck!

A Philosophical Look Back: Tree-Search vs Heuristics as Problem-Solving Techniques

Throughout this book we have been emphasizing what are probably different problem-solving techniques than you were taught and which are not often taught today. These are the five *heuristic* strategies which make up the core of this section of the book. The other problem-solving technique which is generally used in a *tree-search*—a very different way of attacking problems.

Heuristic strategies are broad, general and non-specific.

A tree-search, on the other hand, successively narrows the focus of the questions by choosing the right "branch" of the tree to continue along. As an example, a typical student may very well carry out the following question and answer dialogue unconsciously to arrive at a process for solving a problem.

Q. What kind of problem is this?
A. An adding problem.
Q. What kinds of numbers?
A. Fractions.
Q. Do they have a common denominator?
A. No.

If the student has a good mental "tree" to search and is correct at each branch, this process is quick, efficient and leads immediately to the answer. Some people will argue that the student is not doing a problem, but an *exercise*, that is, a question which he or she already knows how to do. Semantics aside, the students at least finds the answer and learns a process to describe where it came from.

There are, however, two major weaknesses of the tree-search method. First, a tree-search can be *memorized without understanding* where the process connects with anything else. Many students end up learning a large number of mental trees, one for each problem type, which never get assimilated into more general cognitive structures. These trees float more or less at random in the mind without being connected to each other. Eventually, the number of problem types exceeds the students' capacity for remembering trees, and they then become frustrated because their basic strategy for solving problems no longer works.

The second major difficulty is that when a tree search fails, there is nowhere else to go; the students have nothing to fall back on. They are just "stuck." The usual action in these circumstances is to follow blindly along the same tree in the hopes that something will have changed. Such an effort is generally fruitless and does not succeed at bringing the students closer to the answer.

It is here that heuristics comes into its own. For any problem which is not routine, that is, in any situation where the approach is not obvious, heuristics are *far* more useful than a tree search. Heuristic guidelines are almost universally applicable. It does, however, require time and thought to see just which guideline is useful and what the useful application is.

Studying and applying heuristics has another, much greater, benefit, however. It is impossible to apply a heuristic strategy to a situation which you don't understand. The very use of these strategies forces understanding of the underlying mathematical framework. An isolated tree, unlinked to the students' construction of reality, will not remain isolated for too long. Questioning its usage, from a heuristic viewpoint will eventually require the accomodation of that tree into more coherent mental structures. With these more unified views of reality and more general strategies available to them, the students are better able to deal with new and unusual situations.

Good problem solvers, of course, use both types of approaches—often in the same problem. The routine parts of the problem they deal with by a routine tree search. Frequently the difficulties are then reduced to some central subproblem which they can only solve by the application of heuristic strategies. The combination of the two approaches leads to success. Students have been taught tree-search techniques for years and generally can use them well. This is why continual practice in selecting the right (heuristic) strategy is such a valuable skill. And it is why we wrote this book.

We hope your students are more confident, more accomplished and more flexible problem solvers now—in the classroom and in the world.

SECTION IV

Problems to
Illustrate Strategies

Problems to Illustrate Strategies★

Subproblems

Diagrams

Smaller Numbers

Patterns

Working Backwards

Multiple Solutions

Recall that the hundreds' digit denotes the difficulty (100-level problems are easy, 200-level problems are intermediate, and 300-level problems are advanced), the tens' digit denotes the strategy emphasized, and the units' digit the suggested sequence.

The problem variants are rated by stars with (★) indicating that the variant is an easy one up to (★★★), very difficult.

Settling Up the Bills

Three children in the neighborhood decided to throw a party and share the expenses equally. Susan bought $3 worth of ice cream, Janet bought a cake from the bakery for $5, and Sharon bought $1 worth of candy from the candy shop.

How much did each person pay the others so that each of them spent the same amount?

Comments:

Clearly the crucial issue is to decide how much each child should have spent. From here it is a short step to sorting out who owes what to whom.

Hint 1:

How much was spent in all?

Hint 2:

How much was each child's share?

Variants:

(*) Sharon only bought 50¢ worth of candy.
(**) Janet owed Susan 75¢ from the movies last week.

Mr. Lovejoy hired the twins to rake his yard, front and back, for $6. It was a large yard, but the front and back were about the same size. When the time came to start, Bill was not home from his music lesson, so Bryan started alone. Bryan was finished with the whole front before Bill showed up; he had forgotten about the job and stopped at a friend's house. The two of them raked the back together.

How should they split the money?

Comments:

Your students will probably attack the subproblem of "what part of the job did each twin do" in one of the following ways: using fractions $-\frac{1}{2}, \frac{1}{4}, \frac{1}{4}$ or $1, \frac{1}{2}, \frac{1}{2}$ or by using whole number values 2, 1, 1 or $3 + \frac{1}{2}$ ($3).

During the concluding discussion, encourage each group of four to share the methods they used in solving the problem.

Hint 1:

How much of the job had Bryan done before Bill arrived?

Hint 2:

What part of the whole job did they do together?

Variant:

(**) Bryan did half the front. They were paid $7.

My father picked up many bargains at a store that was going out of business. He bought two packages of envelopes, seven picture frames, a basket for my mom, and a whole sackful of cashew nuts. My sister and I made pigs of ourselves. We each ate one fourth of the total nuts in the sack, but there were still 40 left.

How many nuts were originally in the sack?

EVERYTHING REDUCED

PRICES SLASHED

GOING OUT OF BUSINESS

Comments:

The central subproblem here is to answer the questions of the hint. After knowing that, the rest of the problem is easy.

Hint:

How much of the nuts were eaten?

Variant:

(**) Each child ate one-third of the total nuts.

Every afternoon when the two Smith children come home from school, they get some milk and dried apricots. The Smith children like the apricots so much that they eat two two-pound bags every two weeks. Next door, the four Jones children also get dried apricots after school. In four weeks they eat eight two-pound bags.

Does a Jones child or a Smith child eat more apricots in a week? How much more?

Comments:

This problem emphasizes ratios—a terribly important concept. Your students should be able to successively break the problem down into number of pounds of apricots per family per week and then per child. If the arithmetic is too simple, see the first variants.

Hint:

How many pounds of apricots do both Smith children eat in a week?

Variant:

(**) Three Jones children in six weeks eat four two-pound bags.
(**) Change the two-pound bag to a 24-ounce bag.

The Dog, The Goose, And The Bag Of Corn

A poor farmer is going to market with his old and very hungry dog, a plump goose to sell, and a bag of corn. The farmer knows that unless he is right there, either the dog will eat the goose or the goose will eat the corn. He is almost to market when he reaches a small stream, which he must ford, carrying one of the things he has brought with him each time.

How can he get the dog, the goose, and the bag of corn to market safely, and uneaten?

Comments:

The key to this problem lies in finding a good way of representing the various
steps the farmer has to take to cross the river. The easiest way to do this
is to use objects to represent the farmer, dog, goose and corn. Even after
successfully solving the problem, some teams may need help in recording
and/or remembering their solution. This "hidden problem" can and should
be brought out in the concluding discussion.

Hint:

Once an animal or the corn has been safely carried across, it is perfectly
legal to take it back again in order to find the solution to the problem.

Arranging Coins

My sister was fooling around with her money the other night and left the coins in a pattern of four rows with four coins in each row. Each row had exactly one penny, one nickel, one dime, and one quarter; no row, either horizontal or vertical (or even the diagonals), had more than one coin of each kind.

How were the coins arranged?

Comments:

At this point, most of your students should be hunting up objects to represent the coins and sketching a four-by-four square. Leave the diagonal question to those students who wish to try.

Hint:

Maneuvering objects is much easier than erasing "p, n, d, q"

Getting to Grandmother's

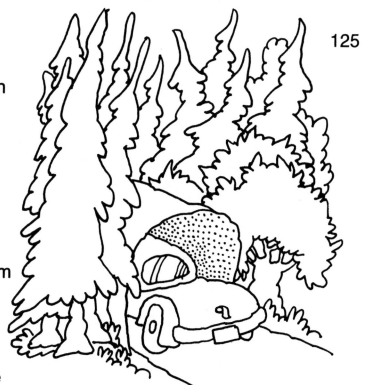

My grandmother moved into an apartment in San Francisco last month, so this weekend my parents drove up from Los Angeles to see her. My mother was driving when we got to San Francisco. I wasn't paying any attention until I heard her say, "I'm lost. Does anyone know how to get to Grandma's house from here?" When I asked what happened, she said that we were at a corner only one block west of the corner where Grandmother's apartment was when we were trapped by the one-way streets. We couldn't just drive the one block. No. First, we had to go three blocks south until we could find a street going east. Then it took us through a tunnel for four blocks until we could turn north. Two blocks north was a dead end. My mother turned west and then asked us for help.

Try to find the most direct route.

Comments:

The key to this problem is deciding to draw a map on which the directions given may be carried out.
Students that have trouble with directions will need extra teacher guidance.

Hint:

Use grid paper to help solve this problem.

Measuring with Sticks

One day while Mark was watching his baby sister, he became bored and started to play with some of the sticks she was playing with. She had quite a collection of all sizes, but would only let him have the green ones, which turned out to be either 5 or 7 inches long. He knew, because it said so on the sticks. There were plenty of both sizes. Then he tried to see if he could make other lengths. Some were easy; if he put a 5-inch stick next to a 7-inch stick, there was a 2-inch gap at one end. He also easily made 10-, 12- and 14-inch lengths. But making one inch baffled him.

Try some experiments to see if you can tell him how to do it.

Comments:

Most students find this problem easy once they can be persuaded to actually draw a picture. Some can do it in their heads, but not many. During discussion, try to get each team to verbalize the methods they used to arrive at a solution.

Note that the question is *how*, not if you can.

Hint:

Would graph paper help?

Variants:

(**)   Use 5-inch and 17-inch sticks.
(***) Use 17-inch and 19-inch sticks.

The Princess's Wedding

For the princess's wedding, the king put on a great wedding feast. He had 34 tables, each with 19 plates of ham. There were 73 tables, each with 26 plates of roast beef, and there were 104 tables, each with 27 plates of vegetables.

How many plates were needed just to serve the ham, beef and vegetables?

Comments:

Students seem able to understand this problem if the panic from the large numbers can be dealt with. Making the numbers easier, in this case smaller, allows them to understand the problem.

Those students who need a calculator and can describe what operations they will use it for should be allowed to use one.

In the concluding discussion, a reference to solving three subproblems is helpful.

Hint:

Suppose there were no vegetables, only two tables each with five plates of ham, and three tables each with four plates of roast beef.

Halloween

The rules of the Witches Guild are very strict. Black hats must be worn in public at all times. Broomsticks are to be replaced yearly and goblin gowns must not contain any patches. On Halloween, each witch must scare 13 people—no more and no less. Each goblin must scare 33 people, and each ghost must scare 19 people.

How many people were scared by a group of 135 ghosts and 273 witches?

Comments:

The only difficulty your students may have with this problem is handling the multiplication, once they understand the question and get rid of unnecessary information. It is assumed that witches and goblins do not scare the same people.

Hint 1:

Suppose witches scare five people each and goblins scare seven.

Hint 2:

How many people would be scared by eight witches and ten goblins?

A Pocketful of Coins

My younger brother wanted a new eraser costing 25¢, so he emptied out his piggybank and got the money he needed. He just had enough to pay for the eraser, even though he had 12 coins.

How many pennies, nickels, and dimes did he have?

Comments:

During the discussion of this problem, students should mention that they
assumed some information about the coins in order to make the problem
easier. With a little experimenting, they should realize that they must have
5, 10, 15 or 20 pennies. Which of these choices will make the problem
easiest? If 20 pennies are used, then only 5 other coins are needed. The
problem is now reduced to the subproblem, can you make 30¢ with 5 coins?

Hint:

How many pennies might he have?

Francisco was raising African spotted mice. They were unusual in that every two months the female had exactly two babies—a male and a female. When these babies were two months old, they had their first babies and continued to have them every two months after that.

Francisco began with one pair of spotted mice on January 1, 1980, and his parents made him sell all that he had on January 2, 1982, just after more babies were born. Since none of the mice had died, he had quite a few.

How many?

Comments:

Hopefully your students will be setting up a chart and looking for a pattern as soon as they understand the problem. Techniques of drawing a picture, using objects, acting it out and/or using smaller numbers may be used.

Hint 1:

What labels on the pattern chart would be helpful?

Hint 2:

How many time periods are there in this problem?

Variant:

(**) He raised the mice until January 2, 1985.

The Yellow Brick Road

In order to advertise their class production of part of "The Wizard of OZ," Mr. Ryan's class decided to put down a small "yellow brick road" around part of their school. The road would be made of sheets of ordinary yellow paper (18 inches by 24 inches) put down as shown in the picture. After a little experimenting, the students also found that they needed to put tape around the outside of all the sheets of paper, either to join sheets together or to stick the paper to the concrete.

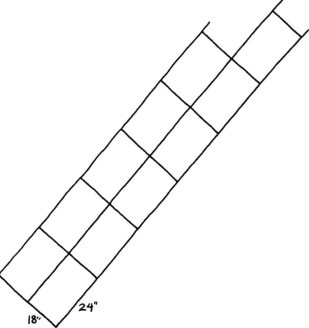

How much tape did they need for a 100-foot-long "road"?

Comments:

The key to this problem is deciding to start with easier numbers and use a chart.

Students may need help in establishing a three-by-two foot measurement per row of paper sheets. You may need to help some groups get started.

Hint 1:

What kind of chart would be helpful in looking for this pattern?

Hint 2:

Let's start with easier numbers. How about one row of paper sheets?

Variant:

(*) Have the road be only one sheet of paper wide.

Playing Investment

We were playing "Financial Management," a new board game. Not long after starting, I had to give up half of my money. Then I had to pay $1,000 in taxes. And then I lost half of the money I had left. By the time this one round of playing was over, I had only $2,500 left.

How much did I start with?

Comments:

After reading the problem silently, many students see that working backwards is the logical place to begin.

Some students may choose to make the problem easier by trying it with smaller numbers first. During the concluding discussion, ask if anyone did this. Whether or not they did, it will provide a nice review.

Hint:

What was the last thing that happened to the Investment player?

The Sale at The Frozen Yogurt Shop

I was walking to school with a group of friends one day last week when Lyn found a flier in the gutter which said that Jake's frozen yogurt store was having a half-price sale on cones that afternoon to celebrate its tenth anniversary of being in business. Everybody got all excited because none of us had heard about the sale before, and we all agreed to tell only a few people (who wants to wait in line forever?).

During morning recess, each of us told one person. Then during the lunchbreak, everyone who knew at that time told one more person. Finally, during the afternoon recess, everyone who knew then told one more person. After school, the 48 people who knew headed directly to the store, and I still had to wait a long time for my cone. But it was worth it.

How many of us were walking to school together?

Comments:

The key to this problem is eliminating the unnecessary information and making a decision to work backwards.

Many students are beginning to find this strategy fun and "cinchie."

Hint:

How many people knew about the sale at the beginning of the afternoon recess?

Mailing the Letter

Claire had just finished writing a letter to her friend, Emma, in England. It had been a long time since she had seen Emma— almost two years. She knew because she had 22 letters that Emma had written her, one for each month since they had last seen each other. This particular letter weighed exactly one ounce, so it would cost 31¢ to mail it. Unfortunately, they were all out of 31¢ stamps in the stamp drawer, so Claire needed to figure out some other combination of stamps which would work. She found a roll of one hundred 15¢ stamps, twenty 4¢ stamps, and five 7¢ stamps.

How can Claire make the exact postage she needs from the stamps she can find?

Comments:

This problem should generate some lively discussion in the groups and possibly some healthy disagreements.

The concluding discussion should include the steps of getting rid of unnecessary information and any assumed information the individual groups might have agreed on. The solution involves solving subproblems and/or finding patterns or recombinations that work.

There are only two solutions.

Hint:

Could Claire use two 15¢ stamps? one 15¢ stamp?

Variant:

(**) Suppose the letter weighed two ounces.

At Blackfoot School, Friday is pizza day, and all of the 60 students look forward to getting their own slice. Mrs. Richards was part way through cutting each of the ten pizzas into six pieces when she discovered one of them had not been cooked. She and Mrs. Hendricks decided to cut some of the pizzas into seven pieces and cut some into eight pieces so that there would be enough pieces to go around.

How many pizzas might have been cut into six pieces? Seven pieces? Eight pieces?

Comments:

Notice that the problem ends with three questions. As students solve the problem in their groups, listen for the various techniques used and bring them to the attention of the class during the concluding discussion. The strategy most often mentioned by my students is solving the subproblem of six, then seven, etc., and exploring the possible combinations.

A question for all to consider: Do there have to be exactly 60 pieces? There are many possible solutions.

Hint:

If Mrs. Richards had already cut five of the pizzas, how should she cut the remaining ones?

Answers:

If exactly 60 slices are desired, the possibilities are:

Solution	6-slice pizzas	7-slice pizzas	8-slice pizzas
#1	6	0	3
2	5	2	2
3	4	4	1
4	3	6	0

(Note the pattern that the solutions form.)

Pancakes for Breakfast

As a special treat for the crew to celebrate their first Fourth of July in space, the cook of the space ship Star Quest is going to make real pancakes for breakfast instead of the usual artificial food which they eat. He calculates that he will need exactly 49 pounds of flour. But the flour only comes in three- and five-pound bags, and he is not allowed to use just part of a bag.

Show how he can get the exact amount of flour that he needs.

Comments:

This problem utilizes the same strategy found in 126. If your students have not solved it, it would be to their advantage to do so.

It combines this type (126) of hidden problem with the one of multiple solutions and many students may set up a pattern to find all the combinations possible.

There are only three possible solutions.

Hint:

How could the cook measure 11 pounds of flour?

Variant:

(**) The flour comes in five- and eight-pound bags.

Answers:

The three solutions are:

Solution	3-pound bags	5-pound bags
#1	13	2
2	8	5
3	3	8

The Hrunkla Village

Captain Spydyk stepped out of his space ship, the Star Quest, on the surface of the alien planet. Hundreds of small creatures resembling furry snakes surrounded his force field while his language-translating computer listened to the strange whistles coming from them. Within about ten minutes, the creatures had arranged themselves into a large rectangle, 24 rows deep with 25 of the creatures in each row, and they were speaking in unison. At this point, the computer was able to translate the key features of what was being said.

It seemed that the Star Quest had landed on top of several of the homes of this village, completely ruining them. All of the creatures of the village made up the rectangle which was talking to the space ship. Most of the creatures were Hrunkla and were arranged in 35 living groups with 17 Hrunkla to a group. The remaining creatures were the Mankla (high rulers), each of whom lived alone. All of the Hrunkla homes were safe, but each of the Mankla needed another home built. It would not take long, but the Star Quest would have to move.

How many homes were needed for all of the Mankla in this village?

Comments:

There are two subproblems: finding the total number of creatures and finding the number of Hrunkla. More extensive comments are included in the
dialogue on pp. 46 and 47.

Hint 1:

How many creatures are there in all?

Hint 2:

How many Hrunkla are there in all?

Variant:

(**) Two of the Mankla homes were not destroyed.

Four students decide to throw a party to celebrate the end of school and share the expenses equally. Amy buys a cake for $8, Randy buys $5 worth of ice cream; Sarah spends $2 on green and red crepe paper to hang, and Dave gets $3 worth of soft drinks. In addition, he pays $6 to rent a giant popcorn popper.

To be fair, who owes money to whom?

Comments:

Clearly the key to this problem is deciding how much each child's share of the expense should be. The process of sorting out who owes what to whom should be shared during the concluding discussion.

Hint 1:

How much was spent in all?

Hint 2:

How much is each child's share?

Washing Windows

Jennifer had been begging her mother all week for some ways to earn money. First, Jennifer cleaned the garage. Then she weeded the garden. Finally her mother agreed to pay Jennifer $3 to wash all the inside windows in the house. Jennifer worked for over two hours, completing 30 windows before her friend, Susan, came over and offered to help her. Each of them washed ten more windows and the job was done. In order to be fair, how much money should Jennifer give Susan for the windows which Susan washed?

Comments:

Usually students try to find the total number of windows washed in all. From this realization they may compute the pay at 6¢ per window and divide up the money accordingly. More advanced students note directly that Susan washed one fifth of the windows and should get one fifth of the money. One group of four, just having completed a unit of equivalent decimals did this: $J = \frac{4}{5} = \frac{80}{100}$; 80 x 3 = $2.40.

Hint 1:

How many windows did Jennifer wash?

Hint:

How many windows were washed in all?

Variant:

(**) Suppose that Jennifer received $4 for the job.

The Class Picnic

Last year, Mrs. Reynolds' class invited the families of her students to come and join in the fun during their annual picnic at Lake Konocktee. Mrs. Reynolds needed to get a white ticket for each adult, a green ticket for each boy, and a yellow ticket for each girl.

The 60 adults mostly sat in the shade watching the kids play. The boys, who made up one third of the total number of people there, generally played in the lake. The girls only made up one sixth of the total number of people at the picnic; most of them played volleyball.

How many yellow tickets did Mrs. Reynolds need to get?

Comments:

Here, careful reading is the key. It leads to the observation that only boys, girls and adults are at the picnic. From there, noting that boys and girls make up half of the people, it is clear sailing to solving the central subproblem: the total number of people at the picnic.

Hint:

What fraction of the people are not adults?

Hint:

How many people at the picnic?

Variant:

(***) Suppose that one-fourth of the total were boys.

New Kinds of Hot Dogs

The "Happy Hot Dog" regional manager was going over the sales figures of the latest new kinds of hot dogs introduced at their seven regional branches. Each branch sold about 500 hot dogs per day. For three weeks, two of the branches had been selling "paprika dogs," while for two weeks the other five of the branches had been selling "southern fried dogs." Since their introduction, 660 of the "paprika dogs" had been sold and 800 of the "southern fried dogs."

Which kind of new hot dog sells better?

Comments:

The key to this problem lies in the students' recognition of the subproblems involved: the number of each kind of hot dog sold each week at each store. These are the numbers needed for comparison.

Although the arithmetic is easy, the students may require some teacher-led discussion to understand the ratio concept.

Hint:

How many "paprika dogs" were sold each week at each store?

In order to demonstrate his peaceful intentions, Captain Spydyk instructed Lieutenant Schwartz, who was an accomplished artist, to prepare a poster or scroll commemorating their arrival on the planet Smygiaa. The message was to be put on a piece of white posterboard 29 inches high and 38 inches wide. It was decided that each letter would be two inches high and that a one-inch space should be left between the bottom of one line of words and the top of the next line. In addition, there should be three-inch borders at both the top and bottom of the posterboard.

How many lines of words could Lieutenant Schwartz fit onto the poster?

Comments:

When a diagram is drawn on and ¼-inch graph paper, it is usually clear to the students that only the height counts and not the width. Then they just fill in the proper squares and count the number of lines. See further comments on page 50.

Hint:

Can you draw a sketch of what the poster will look like?

Variant:

(***) Suppose that a very long proclamation is written on a scroll 38 inches wide and 1289 inches high.

The Circular Track

At our school the track is in a circle with six flags spaced equally around it. Yesterday the three fastest runners in our class lined up at the first flag and started to race all of the way around the track. Elroy took 30 seconds to get to the third flag.

How long did it take him to get all of the way around the track if he kept on at the same rate?

Comments:

This is a problem most students don't recognize. They assume an answer after reading it. Let them deal with it 30 to 60 seconds in their groups and then stop everyone. Tell them that "one minute" is the wrong answer. Ask them to come to group consensus on a different answer and be ready to defend their solution.

Hint:

Most people only find the correct solution after drawing a picture.

Hint:

If stubbornness prevails, require them to evaluate their solution with a picture and/or drama.

The Dog, The Geese, And The Corn

This is an old problem with a new twist. A poor farmer is going to market with his faithful but hungry dog, two plump geese to sell, and three bags of corn. The farmer knows that unless he is right there, the dog will eat a goose or a goose will eat some of the corn. Travelling carefully, he avoids trouble until he comes to a small river which he must ferry across. The ferry (actually a

creaky old rowboat) is in such bad shape, he decides it will only hold him and any two of the six things he has with him.

How can he get all of his possessions to market safely, and uneaten?

Comment:

The key to this problem lies in finding a good way of representing the various steps the farmer has to take to cross the river. The easiest way to do this is to get some small markers to represent the farmer, the dog, the geese and the corn and move them back and forth across an imaginary river. Some students who want a record of what they have done may choose to draw a little river with letters representing each of the actors. (One of our students drew an elaborate set of cartoons showing the rowboat and shores for each trip.)

Hint:

Once the animals or corn have been safely ferried across, it is perfectly legal to take any or all back again in order to find the solution to the problem.

Planting the Orchard

My neighbor has a peculiar orchard. He has five apple trees, five peach trees, five pear trees, five apricot trees, and five plum trees. The trees are planted in a square of five rows of five trees each. Each row (in each direction) has exactly one tree to each variety and so do the center diagonals.

How are the trees planted?

Comments:

Most students immediately begin drawing a five-by-five square and using letter symbols for the trees. This is fine, but they should have a good eraser. Cubes or buttons make it easier to change your mind (but are not necessarily better).

Hint:

Use cubes or buttons for trees.

Variant:

(**) Don't worry about the center diagonals.

The Three Squares

For her Science Fair project, Sarah had found three wooden squares in the closet which she wanted to put everything on. At supper, she was explaining to her family, "I'm going to put the three squares side by side, and then I will put a tape band around the edges of the rectangle that they will form. That way I will use up exactly 120 inches of tape."

"How big are the squares?" her younger brother asked.

"Oh, Billy, you can figure that out", she answered.
Explain the reasoning that Billy had to do.

Comments:

Most students do not read this problem carefully enough to draw the picture correctly. Stress that there is enough information there to solve the problem and help them read it line by line. Note: The intent of the problem is that the tape is continuous, with no overlaps occurring.

Hint:

In drawing the picture, remember that squares are as long as they are wide.

Variant:

(**) Sarah uses four squares to form a large square.

Luis was furious. He was trapped in his bedroom with a broken leg, and his dumb younger brother had run off with the ruler he needed to finish the graph he was trying to draw for his science project. He could use the edge of the picture above his table to draw a straight line, but he needed to measure six inches and he really did not want to go downstairs on his crutches to find the ruler.

Luckily, he remembered his notebook paper is 8½ x 11 inches and he was able to use some sheets of that to get the measurement he needed.

How did he do it?

Comments:

Most students find this problem interesting once they understand the question and realize that actually sketching a few examples is helpful.

They may even add some questions of their own, such as, how many different line lengths could he actually draw with his paper?

Hint 1:

How could Luis draw a line 2½ inches in length?

Hint 2:

How could Luis draw a line 17 inches in length?

Variant:

(**) How could Luis draw a line seven inches long?

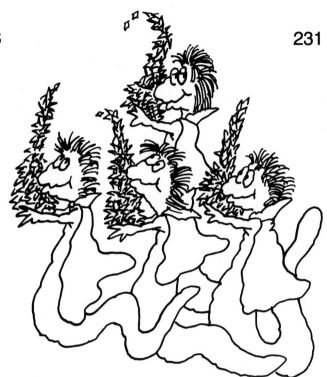

Captain Spydyk looked out at the new part of the alien planet on which his space ship had landed the night before. A huge line of the native creatures were waiting outside the force field protecting his ship. By this time his language-translating computer was able to make some sense of the "speech" and was translating parts of it.

In gratitude for Captain Spydyk relocating the space ship, all 27,804 creatures in the area had contributed to gifts for the crew. As was their custom, each creature had contributed a tiny diamond-like jewel. Pyramids were made out of clusters of 84 of these jewels and lucky individuals in red robes were waiting patiently to give these to Captain Spydyk.

Captain Spydyk worried about the weight involved in accepting so many gifts, even though the jewels looked valuable. But it turned out that a pyramid only weighed 23 grams, so he realized that all of them together would weigh only what?

Comments:

The large numbers tend to frighten students, even though the problem is basically very straightforward. Calculators are a boon on a problem of this sort if your policy permits it. See additional comments on pp. 52-53.

Hint:

Try smaller numbers.

Variant:

(***) Each creature brought 13 jewels.

The Goat Spell

Grizelda the Witch was furious because she was not invited to the inauguration of the mayor of Chicago. So she decided that she would put a spell on all 1,534,464 voters of Chicago to make them think they were goats.

Her recipe read, 'To cast the goat spell, for every 1,728 people, you will need 3 eyes of newts, 4 salamander gizzards, and 2 crystallized dewdrops from a new-grown fern.'

When she went down to the Witches Supply Store, how many salamander gizzards did Grizelda need to get?

Comments:

Students seem to enjoy this problem but still get nervous at the size of the numbers. If they simplify by assuming there are 1,000 people and the recipe is for each 5 people, then the problem becomes quite straightforward.

Hint:

Which information is necessary? How can the strategy of making numbers easier help in finding the solution?

Variant:

(**) Suppose there are 1,535,420 voters in Chicago?

General Wholesale Grocers has just purchased 4,820 cartons of tomato paste which it wishes to stack in its warehouse. Because the cartons are so small, the foreman at the warehouse plans to stack them in a rectangle which is 13 rows wide and 22 rows long on the floor. If he carries out his plan, how many cartons high will he need to stack the tomato paste? (There will be a fraction and/or remainder in this answer. Think carefully about what this means.)

Comments:

Many students already have good estimation skills and round each number to the nearest ten or thousand. The difficulty comes in agreeing on the proper solution because the quotient has a remainder. Most students eventually decide they must "round up" even though the remainder is less than half. In fact, only one carton left over would require a new level to be started.

The arithmetic is also a problem. Our school district allows calculators in the classroom. If a team can write a headline and requests use of a calculator, our policy is to let them use one. This is our preference; you must do what is acceptable in your district and what you are comfortable with. *Remark:* Problems 133 and 333 are not very similar to this one.

Hint 1:

What if there were only 100 cartons put in 10 rows, 5 rows wide?

Hint 2:

What if there were 101 cartons in the same arrangement?

The Creature from the Black Lagoon

A rare earthquake in Florida allowed a strange creature who had been trapped underground for centuries to come to the surface and begin to grow again. It would swallow any plant whole and grow and grow and grow. Each day it doubled in size. At 8 A.M. on August 14, it was first seen by fishermen who estimated its size to be a blob of about 100 cubic feet. For five days it grew, and nothing would stop it until finally at 8 a.m. on August 19, it slipped into the Atlantic Ocean and died of salt-water poisoning.

How big was it when it died?

Comments:

One or two groups usually arrive at the same incorrect solution, i.e., 5 x 124 = 620 cubic feet. Although this has a discouraging side, it presents a wonderful opportunity to utilize the strategy of fewer steps. What is his size after two days? What happens next? What is his size after three days . . . etc.?

It also reestablishes the techniques of checking and/or evaluating the solution with the problem.

Hint 1:

Make the problem simpler—use fewer days or a smaller size for the creature or both.

Hint 2:

What size was he on August 16, at eight a.m.?

Variant:

(**) The creature did not die until August 29.

●   My little sister decided that she just had to have a new softball. So she went to the bank where she hoards her money and emptied it out completely in order to get the 90¢ needed for the ball. The money was in the form of 45 coins which made her purse a little heavy for her to carry, and it also took a long time for her to count out the money at the store. (I had to take her, and I got bored waiting.)

How many pennies, nickels, dimes, quarters and half dollars did she have with her?

●

●

Comments

During the init discussion of this problem, students should mention the
assumed inforr ion, i.e., she did have some pennies, and she did not
have a half dolla

At the conclud discussion, emphasis should be on the methods used
to solve the probl How did the "making numbers easier" strategy help
in solving the prob n? As this problem has three possible solutions (all
correct), the discus n will be lively.

Hint:

How many pennies mi she have?

Handshake

● If everyone in your class shakes hands with everyone else, how many handshakes would there be?

●

●

Comments:

This problem gives rise to many patterns detail d on pp. 55-56, some being more useful than others for answering the origi__ question.

Hint:

Set up a pattern.

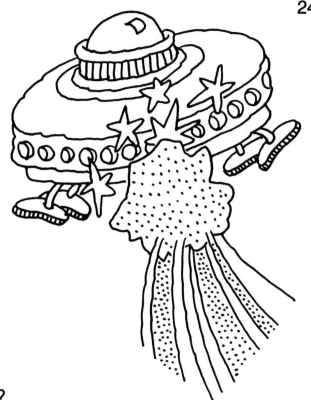

About ten million light years from the center of the Milky Way galaxy is the Intergalactic Hilton, the finest and most up-to-date of all the intergalactic hotels. Each of the 54 species of creatures which uses it has its own living accommodations in a spinning wheel designed to approximate the gravity on its home planet. From the center of each wheel runs a message tube to the center of each other wheel.

How many such tubes are needed?

Comments:

Groups may have trouble getting started—if so, see the hint. There may be comments from the groups of students about using easier numbers and/or drawing a picture. These strategies are helpful in gettting started. Once the work is underway, however, the goal is for the students to see the pattern and use it to solve for 54 wheels.

Many of these pattern problems can be written as equations (functions) and if you wish to encourage this, please do so.

Hint:

Let's start with easier numbers and set up a chart . . .

	Wheels	Tubes
How many tubes with 1 wheel?	1	0
2 wheels?	2	1
3 wheels?		

Teaching Reading

In the newly developing country of Zamwana, teaching reading is a focal point of the government program. The plan is for each individual who can read to spend one year teaching two others to read. That individual is done, but the two new readers must spend one year, each teaching two others to read.

If 1,000 people can read in Zamwana now, how many people will be able to read in ten years?

Comments:

Hopefully your students will be setting up a chart looking for a pattern as soon as they understand the problem. Techniques of drawing a picture, using objects, acting it out and/or using smaller numbers may also be used.

Check to be sure the students understand that each reader only teaches for one year.

Hint:

Rely on the strategy of easier numbers. Only one person knows how to read—how many people will know how to read in five years?

Variants:

(***) Everyone who can read keeps on teaching each year.

My uncle has been sending me a lot of books over the past few years, all of them Nancy Drew mysteries, which I love. Depending on his mood, I may get anywhere from one to six books. During this time I have been keeping track of the postage he has had to spend. (During this whole time the postal rates have been the same). A package with two books costs 85¢ to mail; one with five books cost $1.75. A package with one book costs 55¢; one with four books cost $1.45; and each package of three books cost $1.15. How much did it cost him to mail a package of six books?

How much would it have cost him to mail all 52 books at once?

Comments:

The first question is easily answered by listing the prices and number of books sequentially. The second question becomes simple once the pattern is identified. Although it's a two step function, the math in this problem is quite simple and students seem to deal with it successfully.

It is explained quite nicely in words so that insecure students will not have to deal with an equation to feel successful. The discussion period on this problem is usually a comfortable time to write the pattern (function) both ways.

Hint:

What kind of chart would be helpful in looking for patterns?

Number of Books	Price

Golden Apples

A prince picked a basketfull of golden apples in the enchanted orchard. On his way home, he was stopped by a troll who guarded the orchard. The troll demanded payment of one-half the apples plus two more. The prince gave him the apples and set off again. A little further on he was stopped by a second troll guard. The troll demanded payment of one-half the apples he had then plus two more. The prince paid him and set off again. Just before leaving the enchanted orchard, a third troll stopped him and demanded one-half of his remaining apples plus two more. The prince paid him and went sadly home. He had only two golden apples left.

How many apples had he picked?

Comments:

This is an excellent example of a problem which is most easily solved by working backwards. Suppose that the prince had met only one troll; and under these circumstances how many golden apples would he have had? See additional examples on page 58.

Hints:

How many golden apples did he have before he met the third troll? Before he met the second troll?

Variant:

(**) Each troll demands half his apples and one more.

The Shopping Trip

My brother, Michael, loves to go shopping. As soon as he gets his clothing allowance, he heads for the nearest store. Yesterday he got his usual amount of money and went to buy jeans. In that store, he spent half of his money and three dollars more. Then he went to buy a shirt and in that store, he spent half of his remaining money and two dollars more. After that, he had only five dollars left.

How much did he start with?

Comments:

After reading the problem silently, many students see the similarities between this problem and the Golden Apples. After restating the problem and questions, the teams are ready to go to work. (Ask them what strategy they plan to use.) During the concluding discussion, you may be surprised by the solution one team gets. Usually one or two teams will not always check their solution with the stated problem. If one of these teams also does not read the problem carefully, they get caught. One team, thinking all working backward problems are alike, used the $5 amount stated at the end of the problem as the amount he had left over after each purchase. That was the last time the team forgot to check their solution.

Hint:

How much money did he spend on his last purchase?

Variant:

(**) In the second store, he spent one-third of his remaining money and two dollars more.

Lost in the Desert

My best friend is a desert nut, and she persuaded me to go out and hunt thunder eggs (or geodes, rock balls from volcanic eruptions) with her one Saturday. Her father dropped us off and agreed to come back in six hours. Carol led the way, and we had a good time until she realized that we had left our compass and map in the car. Then we decided to head back to the highway but got lost.

We did not have that much water and drank half of it the first day. The second day, we were only going to drink half of what was left, but we drank that and a cup besides. When the search party found us that evening, we had only two cups of water left. I'm *never* going out with her again.

My mother figured out how much water we started out with, but I didn't care. I just don't want to see the desert for at least ten years.

What was my mother's answer?

Comments:

The keys to this problem are eliminating the unnecessary information and making a decision to work backwards. Many students are enjoying this strategy by this time and think the problem a "cinch"! A concluding discussion might include instances in real life when this strategy would be helpful.

Hint:

How much water did the girls drink the second day?

Freaking Out

Bill, Nancy, and Fred had been restless for two hours during their visit to Aunt Sophie, so Mother finally gave each of them 50¢ to spend at the dime store up the street. The three children had their coats on and were out of the door in less than a minute and ran almost the entire way to the shop. Fred walked the last block because, he said, his side hurt. When they got there, they had their choice of plain erasers (three for 10¢), colored pencils (5¢ each), animal-shaped erasers (two for 25¢), or a small pad of colored paper (15¢ each). Each child bought at least two things; each made a different choice; and each spent all of his or her money.

Can you tell what they could have bought?

Comments:

This problem should generate some lively discussion and possibly some healthy disagreement within each group of four.

Various techniques such as organizing the needed information need to be utilized before the various subproblems and/or patterns of purchase can be figured out.

There are 20 ways to spend 50¢ under these conditions.

Hint:

If I bought two animal-shaped erasers, what else could I buy?

Solution	Plain erasers	Pencils	Animal erasers	Paper
#1	0	1	0	3
2	0	4	0	2
3	0	0	0	2
4	0	2	2	1
5	3	2	0	1
6	0	7	0	1
7	3	5	0	1
8	6	3	0	1
9	9	1	0	1
10	0	5	2	0
11	3	3	2	0
12	6	1	2	0
13	3	8	0	0
14	6	6	0	0
15	9	4	0	0
16	12	2	0	0

The Weber City Boy Scout Troop #15 was on its annual campout at Fallen Leaf Lake. The scoutmaster had arranged for 20 boats to be available at the dock to carry the 120 campers across the lake to the campground. However, when they all arrived at the lake, they found that some of the boats were not there. So instead of each boat carrying six people, some of the boats carried six, some carried seven and some carried eight.

How many boats might have carried seven people?

Comments:

Notice the use of the word *might*. This should clue in most students that there may be more than one possible solution. In order to answer the question, students may need to solve the hidden problems: how many might have carried 6? 8?

There are many possible solutions as the number of boats is unknown.

Hint:

If there were 18 boats, how might the scouts have been transported?

Mixing Feed in Nigeria

● The uncle of one of my
classmates came to our class
recently and told about his
experiences in teaching
agriculture in Nigeria. The country
has recovered from its civil war of
a decade ago and is trying very
hard to improve its food output.
This man was one of the 85
people from U.S.A.I.D. who were
working with the Nigerians trying
to help. At one point, he was
showing some of the farmers how to mix some corn with the cows' usual
food and wanted to weigh out 90 kilograms of corn for all of the cows.
Although there were no scales, there were two empty kerosene tins
● available. The smaller one he knew would hold five kilograms; the larger one,
8 kilograms.

   Then he asked us how he could have measured the 90 kilograms needed.
It took us awhile to figure it out. Show how it might be done.

●

Comments:

This problem utilizes the same strategy found in 226. If your students have not solved it, it would be to their advantage to do so.

It combines this type (226) of hidden problem with the one of multiple problems containing multiple solutions, and many students may set up a chart and/or pattern to find all of the possible combinations.

There are only three possible solutions.

Hint:

How could he measure 21 kilograms of corn?

Four students decide to throw a party to celebrate the end of school and share the expenses equally. Amy buys a cake for $8, Randy buys $5 worth of ice cream, Sarah spends $2 on green and red crepe paper to hang, and Dave gets $3 worth of soft drinks. In addition, he pays the $6 needed to rent a giant popcorn popper.

To be fair, who owes money to whom?

Comments:

Clearly the key to this problem is deciding how much each student's share of the expense should be. The process of sorting out who owes what to whom should be shared during the concluding discussion.

Hint 1:

How much was spent in all?

Hint 2:

How much is each student's share?

Variant:

(**) Pat joins the group at the last minute and only brings $1.50 worth of peanuts.

●   In my home town, there aren't
many ways to make money, but I
deliver advertisements for a local
supermarket every week. They
give me 455 of them to deliver
and pay me $2\frac{1}{4}$¢ each to do so.
Usually, I give Karen, my fourth-
grade sister, $2 to fold all of the
ads and to put rubber bands
around them for me. Yesterday,
however, after she had been
folding for an hour, my aunt and
her six-year-old son, Brad, showed up. Brad wanted to help fold. Karen
agreed, and the two of them began working. Brad tried hard, but he could
only take care of half as many papers as Karen. After they had worked
●   together for an hour, all of the papers were done.

   To be fair, how much of the $2.00 should I pay to Brad.

●

Comments:

This problem contains unnecessary information. Nevertheless, the students should realize that Brad effectively did as much work as Karen in half an hour, that is, one-fifth of the total. It is also quite possible to calculate that Karen folded 364 to Brad's 91 ads, then to make the computation that way.

Hint:

How long would it have taken Karen working alone?

Variants:

(**) There are 400 papers to fold.
(***) Brad works one third as fast as Karen. [In this case, the "correct" answer will involved a fraction of a cent.]

Eating the Mints

After Thanksgiving dinner, Grandmother put a large tray of mints on the coffee table by the TV for people to munch on while they watched the football game. Uncle Frank loves mints, and he managed to eat half of them during the game. My father almost did as well as Uncle Frank; he ate one third of all the mints. Grandfather isn't supposed to have any mints, but he took one anyway, and I ate the remaining three.

How many mints did Grandmother put on the table originally?

Comments:

Students should come quickly to the realization that the central subproblem is find the fraction of mints not eaten by Father or Uncle Frank. They will then be able to go immediately to the answer from there. (This assumes that students understand equivalent fractions and can add them.)

Hint:

What fraction of the mints were eaten by Father and Uncle Frank?

Variant:

(***) Father and Uncle Frank each ate one third of the mints; Mother ate one fourth.

Washing Hair

Two women were complaining about their teenagers' preoccupation with clean hair. One moaned, "My four boys will use up three bottles of shampoo in two weeks." The other replied, "My five girls will use up four bottles of shampoo in three weeks." Who uses the most shampoo in a week, one of the boys or one of the girls? How much more?

Comments:

The key to this problem lies in the students' recognition of the subproblems involved: How many people use how many bottles of shampoo each week. These are the numbers needed for comparison.

Although the arithmetic is not difficult, the ratio concept may need some helpful teacher input.

Hint:

How many bottles of shampoo did each boy use each week?

Capping the Bottles

Last month our class took a field trip to a local bottling plant to see how they make Kalo-Kola. We watched everything from pumping the sugar syrup out of tank cars to putting the tops on the bottles. Everyone commented that the machine looked just like the one at the beginning of the old "Laverne and Shirley" show—a circular device holding 36 bottles equally spaced around the outside.

When I saw it, I tried to figure out how long it would take to cap all 36 bottles. I started timing right after the top was put on the first bottle and found that it took 10 seconds until the ninth bottle was capped. Then Mr. Tolving said it was time to head back to school.

How long would it have taken for the machine to cap all 36 bottles?

Comments:

This is a problem many students won't recognize as such. They will assume an answer after reading it. Let them deal with it for 30 to 60 seconds in their groups and then stop everyone.

Tell them that "40 seconds" is the wrong answer. Ask them to come to group consensus on a different answer and be ready to defend their solution.

Hint 1:

Most people only find the correct solution after drawing a picture.

Hint 2:

How many bottles were actually capped in ten seconds?

Crossing at White Falls

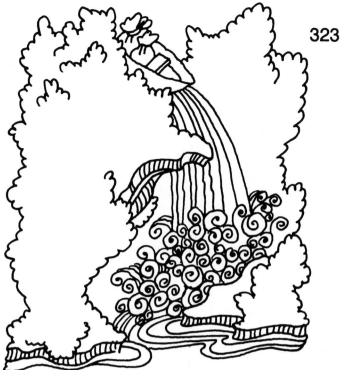

It was during the great gold rush in the Klondike that this happened to Grandfather Bejarnik. He was hired to guide ten men with their gold out of the MacKenzie River country and, as it turned out, to keep them from stealing from each other. Any of the four Harrison brothers would gladly have taken the gold from the three Johnsons, but not from either of the two Smiths or from Hank Rivers, who were all their cousins. Any of the Johnsons, on the other hand, would steal from the Smiths or Hank Rivers; either of the Smiths would steal from Hank (they seemed to have no family feeling). Unless my grandfather was right there, he knew something would happen.

All went well until they reached White Falls crossing, where the only way to the other side of the river was in a rowboat which would hold five men at most. Nonetheless, my grandfather managed to get everyone across the river without any group being alone with another group (or person) whose gold they wanted. Can you tell how?

Comments:

This is our old friend about the dog, the goose, and the bag of corn decked out in new finery. Counters are good for this problem, but encourage your students to keep some record as the solution is quite long. If the groups are having a hard time getting started point out that men from more than one family can ride in the rowboat at once.

Hint:

Who are the only four men that Mr. Bejarnik can take across the river so that the gold of those remaining is safe?

The Temple's Cube

Lieutenant Curtis reported back to Captain Spydyk on their visit to the seat of the government of the Hrunkla. The king sat upon a giant cube which was broken down into 64 smaller cubes all the same size (consisting of four layers of cubes each one four by four). There were 16 white cubes, which looked like alabaster; 16 black cubes, which looked like onyx; 16 green cubes, the color of jade; and 16 deep red cubes, which seemed to have no equivalent on Earth.

It was asserted that each vertical direction and each horizontal direction, through the cube had exactly one small cube of each color. Spydyk doubted this, but Curtis showed him how it might be done.

How?

Comments:

Students seem to need more than objects (cubes) to solve this problem. They appear to need experience. If they have not done #224, have them do it first. The same techniques can be applied to this three-dimensional problem. Notice that no mention is made of any diagonal.

Hint 1:

What might the top layer look like?

Hint 2:

Once you have the top layer, can you see an easy way to get the next layer?

Answer:

For top, take any suitable horizontal layer such as

```
W  B  G  R
R  G  B  W
B  W  R  G
G  R  W  B
```

For the second layer replace W by B, B by G, G by R and R by W to get

```
B  G  R  W
W  R  G  B
G  B  W  R
R  W  B  G
```

Continue replacing colors in the same way to get the third and fourth layers at right.

Third				Fourth			
G	R	W	B	R	W	B	G
B	W	R	G	G	B	W	R
R	G	B	W	W	R	G	B
W	B	G	R	B	G	R	W

Hrunkla Apartment Houses

Most Hrunkla lived in giant twelve-story apartment houses, and their homes were large square rooms bounded on four sides by corridors. Each room had a single door which opened half way along a corridor. On even-numbered floors, the doors opened onto the east corridor; on odd-numbered floors, the doors opened onto the north corridor. At each intersection of corridors,

there was something like an elevator which could be ridden up or down.

Half of the corridors had moving belts on the floor, and no self-respecting Hrunkla would walk if he could ride one of these belts. The belts were so arranged that those on floors 1, 5 and 9 ran to the east; those on floors 2, 6 and 10 ran to the south; those on floors 3, 7 and 11 ran to the west; and those on floors 4, 8 and 12 ran to the north.

Describe how Hrunkla who lived on floor 10 could use these moving belts and elevators to visit a friend who lived in the room directly below his.

Comments:

The key to this problem is the ability not only to see the necessity of drawing a diagram but to see that this problem consists of several subproblems—the corridors, doors and moving belts on various floors.

Hint 1:

Where are the doors and corridors on floor 10? On floor 9?

Hint 2:

What direction does the belt move on floor 10? On floor 9?

Answer:

There are many correct ways. The shortest one is to ride the belt on floor 10 south to the corner. Ascend to floor 12 by the elevator. Ride the belt on floor 12 north to the corner. Descend to floor 11 by the elevator. Ride the belt to the door.

Cooking Rice

Two weeks ago we went on a
camping trip to Eagle Lake, a
place I had never seen before. It
was a beautiful place, and our
campsite was quite near the water.
At suppertime, we decided to
make rice. We started to cook the
rice for everyone. It had been
measured at home so we knew
that we needed exactly five cups
of water if it were going to be cooked correctly. Unfortunately our camping
set did not have a pot which held exactly five cups. We did have one pot
which held exactly seven cups and another which held exactly three, but I
couldn't see any way of measuring five cups. Luckily my mother is smarter
than I am, and she figured out a way.

Can you see how she did it?

Comments:

Most students find this problem interesting once they understand the question and realize that sketching a few examples is helpful.

They may even add some questions of their own, such as how many different amounts can Mother actually measure given these two pots?

Hint 1:

How could Mother measure four cups?

Hint 2:

How could Mother measure six cups?

Answer:

Fill the 7-cup pot from the lake, Pour from it into the 3-cup pot. Empty the 3-cup pot. Fill the 3-cup pot again from the 7-cup pot. Empty it again. Pour the remaining cup into the 3-cup pot. Fill the 7-cup pot again from the lake. Pour as much as possible (2 cups) into the 3-cup pot. The larger pot now holds 5 cups of water.

Intergalactic Conference

At the intergalactic conference, the space problem was terrible. Each of the 1,328 Smurgians required .341 cubic meters of space, while the 139 Zaargs each needed 2.46 cubic meters, and the 417 Xxegs insisted that each of them needed as much as a Smurgian and a Zaarg together.

How much space was left over if the building held 2,100 cubic meters?

Comments:

Students may panic at the use of decimals, cubic meters and large numbers. Making the numbers easier, in this case into whole numbers and/or multiples of hundreds, allows them to understand the problem and solve it.

In the concluding discussion, a reference to solving three subproblems is a helpful review.

Hint:

There are 100 Smurgians—each needs three cubic meters of space. How much space do they need all together?

Driving Across the Desert

I don't know if you have ever driven across Nevada, but it can get pretty boring. There are miles and miles of nothing but miles and miles. I was riding with my cousin who had set the cruise control on his car so that it travelled at a steady speed. For one five-milestretch, I timed us on my wristwatch and discovered that we had taken exactly four minutes and thirty seconds. Just then we saw a sign, "McDermitt, 73 miles."

How much longer did it take us to reach McDermitt?

Comments:

The problems encountered here are dealing with minutes and seconds and an odd number of miles. During the concluding discussion, ask the teams what methods of making numbers easier they used to solve the problem.

Hint:

Suppose it took five minutes to go five miles. How long would it take to get to McDermitt? In the original question is the car going faster or slower than the one in this hint?

Variant:

(**) It is 70 miles to McDermitt.

Protozoa

These tiny creatures are probably the fastest growing things in the world. If you put some suitable protozoa in a culture where the temperature and humidity and food available are just right, they can increase in weight 200% in just four hours. This rapid increase goes on until they run out of food.

Suppose you had one gram of such a protozoa in a perfect situation for it to grow as rapidly as possible and enough food for it to grow for two full days.

How many grams of protozoa would you end up with?

Comments:

An initial discussion may be necessary to clarify the meaning of an increase in weight of 200%.

Once this is clear, the process now becomes one of utilizing the strategy of fewer steps. What is the size after four hours? What happens next?

In the concluding discussion, you may wish to review the technique of checking the solution with the problem—day by day.

Hint:

How many grams of protozoa do you have after four hours?

Mother's Day

For Mother's Day, Davie, my little brother, Kathy, my younger sister, and I all contributed money to buy a present for Mom. Davie had saved 80 pennies, two nickles, and one dime. Kathy gave me three half dollars which she had saved, and I contributed the rest. Actually, with what Davie and Kathy gave me, the 17 coins in my bank were just enough to make up the total cost of $8.12.

What coins were in my bank?

Comments:

During the initial discussion of this problem, some assumed information needs to be clarified. A valid question concerns the silver dollar; the solution is possible either way and the option of leaving it alone is valid, also.

   In the concluding discussion, emphasis should be placed on the methods and/or steps used to solve the problem.

Hint 1:

What denomination of coin did we have for sure—at least how many?

Hint 2:

By breaking the problem down into fewer steps or easier numbers, how can you start to solve the problem?

Sumatran Lilies

Sumatran lilies grow from bulbs, very much like tulip bulbs except that at the end of the summer, the original bulb has produced two more. During the next year, each of the three bulbs will do the same. On a small island off of the coast of Sumatra, a determined farmer is attempting to grow as many of these bulbs as he can in ten years of growing and then sell all of the bulbs and retire.

If he begins with 73 bulbs, how many does he have to sell when he retires?

Comments:

Hopefully, by this time your students will be setting up a chart and looking for a pattern as soon as they understand the problem.

If your students have not studied exponential growth, this pattern can serve as a nice introduction.

Hint:

Make the problem simpler by having the farmer begin with only one lily.

Variant:

(***) If each bulb weighs 2 ounces, how many 100 pound bags of bulbs will the farmer have?

Captain Spydyk's Bank Account

During the years he is in space, Captain Spydyk has a special bank account with $20,000 put into it directly at the beginning of each year. The bank pays him 100% interest on his money as a special favor, so that whatever money he has in the bank at the beginning of the year is doubled at the end.

If Captain Spydyk stays in space for ten years, how much money will be in his account?

Comments:

The key to this problem is deciding to look for a pattern and setting up a chart.

Many students will find the pattern easier to see if they drop the last three or four zeroes from each year-end total. This pattern can be written as a function and should provide some interesting learning during the concluding discussion.

Hint:

What kind of chart would be helpful in looking for a pattern?

Years	Total

The Coat on Sale

● Harris's Department Store was trying to sell a winter coat which no one seemed to want. First, they tried to sell it for 20% off the marked price, but no one would buy it. Then, they tried to sell it for 25% off the first sale price, but still no one would buy it. Finally, they offered it for 30% off of the second sale price, and someone actually bought the coat for $21.

What price did Harris's originally try to sell the coat for?

Comments:

There are two common ways to do this problem. One is working backwards, calculating that the second sale price was $30, the first sale price $40, etc. The other approach is to observe that the first sale price is $4/5$ of the original cost, the second sale price is $4/5$ x $3/4$ of the original cost, and the final sale price $4/5$ x $3/5$ x $7/10$ = $21/50$ of the original cost.

Hint:

How much was the coat on sale for during the second sale price?

Variants:

(**)   Suppose the first discount was 30%, the second 25% and the third 20%

(***) The first discount is 20% plus $4; the second discount is 25% plus $3; the third discount is still 30%.

The Bears

Last summer we went camping in Yosemite, and the first night we did a dumb thing: we left our food on the ground. A bear came along and ripped up one third of our total number of dried meals. Next day we ate four more of the meals and tied the food up in a tree. It didn't seem to help because one third of the meals we had left were ripped open by another bear. During our third day, we ate four more meals and that night, despite everything we did, one half of our remaining dried meals were ripped apart.

We gave up, ate the four remaining dried meals, and headed home.

Can you tell how many dried meals we started with?

Comments:

This problem is almost impossible to solve without working backwards. The major stumbling block seems to be in working backward from the morning of the third day (when 12 dried meals remained) to the evening of the second day. To do so the students need to realize that 12 represents 2/3 of the meals left after the second evening's meal—that is, 18. Going backwards from the morning of the second day to the evening of the first is similar.

Hint:

How many dried meals were left after the supper on the third night?

After the Volleyball Game

Our volleyball team had just finished playing in a tournament, and the 20 of us decided to go to the yogurt shop across the street for a snack before heading home. Each of us ordered either a small cone (for 50¢), a large cone (for 70¢) or a large dish with nuts and carob chips (for $1.30). After everyone had ordered, the bill came to exactly $19.00. On the way home, I was trying to remember what people had ordered, but I finally had to give up. I did decide that more than five people had ordered large cones.

How many people might have ordered large dishes?

Comments:

This is a good homework problem, best solved by making some calculations about the number of people who ordered large dishes and going from there. If only eight people ordered large dishes, then even if everyone else orders a large cone, the total cost is less than $19. On the other hand, if twelve people order large dishes, and everyone else orders a small cone, then the total is too high. So only nine, ten or eleven people can have large dishes. Each of these cases may be treated as a separate subproblem.

There are only three solutions.

Hint 1:

Suppose five people ordered large dishes. What is the largest possible bill in this case?

Hint 2:

Suppose fifteen people ordered large dishes. What is the smallest possible bill in this case?

Mr. Rhodes said to his two sons, "Jim and Paul, I have a load of 200 bricks in front of the house to put around the edge of the patio, and I want you two boys to carry them around to the back of the house this afternoon. I find that eight bricks is about right to carry on each trip."

After they had hauled a few loads, Jim said, "I think we could be done faster if we would carry nine or ten bricks on each trip." Paul agreed and that is what they did for the remaining trips.

How many trips might they have made with eight bricks? with nine bricks? with ten bricks?

Comments:

There is some deliberate ambiguity in the statement of the problem: how many tries are "a few"? Some students are bothered by this imprecision. It is also left vague as to whether or not both boys make the same number of trips. Either assumption may be made, but your students should be clear on what they take the problem to mean.

There are many correct solutions to this question.

Hint:

Suppose Jim made three trips and Paul made two trips with eight bricks each, how many bricks are left? Find two ways to carry the remaining bricks around.

Variant:

(***) Suppose there were 225 bricks?

Answer:

Suppose that "a few" means 2 to 6 trips. Then the possible solutions are below.

Solution	Trips with 8 bricks	Trips with 9 bricks	Trips with 10 bricks
#1	6	8	8
2	5	10	7
3	4	12	6
4	3	14	5
5	2	16	4
6	5	0	16
7	4	2	15
8	3	4	14
9	2	6	13

Notice the two separate parts, each with its own pattern.

Lining the Field

You are manager of the football team, and your job this afternoon is to put the lines on the field for the game on Friday. Everything is easy except for the fact that you are supposed to put the hash marks one third of the way across the field (which is 160 feet wide). All you have at your disposal are the chain markers, exactly 10 yards long, a five-foot hash marker outline, and a 40-inch pointer which the coach sometimes uses.

How is it possible to locate the lines?

Comments:

Once the hint is accomplished, the problem becomes finding 53 1/3 feet using a 30-foot, a 5-foot and a 3 1/3-foot measure. Phrased this way, the situation rarely gives trouble.

An occasional very bright student may use 4 inches as the basic unit of length and reduces the problem to finding 160 of these units with 90, 15 and 10 unit measures.

Encourage your students to draw a picture.

Hint:

Can you convert all of the measurements to the same scale (yards, feet or inches)?

Variant:

(****) The coach's pointer is 44 inches long.

Answer:

There are nine possible ways without measuring backwards.

Solution	Chain	Hash marker	Pointer
#1	1	0	7
2	1	2	4
3	1	4	1
4	0	10	1
5	0	8	4
6	0	6	7
7	0	4	10
8	0	2	13
9	0	0	16

Mind-Expanding Problems

Mind-Expanding Problems*

Easy

Medium

Difficult

---

*The numbers in parentheses after the title of the problem refer to strategies that may be useful in solving it:
- 0. Techniques for Understanding the Question
- 1. Subproblems
- 2. Diagram, Manipulatives
- 3. Simpler Numbers, Estimation
- 4. Patterns
- 5. Work Backwards

Going to the Movies

Jay and Chris went to the show. They left the house at 6:45 p.m. They arrived at 7 p.m. Chris bought their tickets and waited in line for 22 minutes. She got $4.64 in change from her $10 bill. When they finally got inside the theater, Jay stood in line to buy popcorn. He was tenth in line. He bought two tubs of popcorn, each at 85¢. Halfway through the movie Jay went back to the snack bar to get two orange drinks and another tub of popcorn. Much to his surprise, one drink cost 75¢. How much money did Jay spend on popcorn?

Comments:

This problem is included solely as an example of a problem with unneeded information. (see p. 62).

As Bill was heading off for the Little League diamond his mother tossed him the tape measure saying, "Don't forget that we are supposed to cover the shelves at the snack booth with paper. Find out how much paper we will need." When Bill finished practice, he measured the shelves. Four of them were eight feet long and ten inches wide while the other two were seven feet long and eleven inches wide. Write down what Bill told his mother.

Comments:

This problem is included solely as an example of a hidden question (see page 37).

Spydyk's Birthday

The space ship, Star Quest, is beginning its sixth month in space with a celebration of Captain Spydyk's birthday at 1800 hours. Alice and Mark were in charge of decorating the main control room for the party as it was the largest room on the ship. To make more room, they were moving out three supply boxes which each weighed 80 pounds on earth.

Alice told Mark she would be there 45 minutes early so they could get everything done on time. Mark wrote the time on a slip of paper which he put into is pocket. Later, he decided to be there 20 minutes before Alice to get the supply boxes moved.

He arrived just when he wanted, moved the boxes and then helped Alice decorate. It was a wonderful party which everyone thoroughly enjoyed. But Mark could never remember when he arrived. When *did* Mark arrive?

Comments:

This problem is included solely as an example of a problem with much
irrelevant information (see page 37).

The Wedding Gifts

When the mean, nasty princess of Ruritania was married, the prince from the neighboring kingdom of Walpurgis was so relieved that he had not had to marry her that he sent the newlyweds five dolls made of silver and dressed in the costume of the country for the princess's collection. Each doll was two ounces heavier than the preceding one.

If the heaviest doll weighed 24 ounces, how much did the dolls weigh in all?

Comments:

The solution to this problem may involve three strategies: pattern recognition, working backwards and the subproblem of addition. Many students find the pattern table helpful.

Dolls	Weight
1	
2	
3	
4	
5	24

Hint 1:

Make the problem easier by having him give her only two dolls.

Hint 2:

If doll #5 weighs 24 oz, what is the weight of doll #4?

Variant:

(*) Ten dolls were given.

Elk Grove School District has just hired a music teacher, Mr. Leffler, to handle the music programs at its three elementary schools. On Monday and Wednesday he teaches at Anthony and Johnson schools; on Tuesday and Thursday he teaches only at Central; on Friday he teaches at Anthony and Central.

Mr. Leffler lives only three blocks from Central school, so he always walks there and home again. However, he has to drive 2.7 miles to get to Anthony school from his home and he has to drive 1.2 miles to get to Johnson from his home. To get from Anthony to Johnson he has to drive 1.9 miles.

How many miles will he have to drive for his teaching alone if he will be teaching 30 weeks this year?

Comments:

This problem can be solved in a series of steps, often referred to as sub-problems. Some students choose to solve this problem by grouping sets of days with like travel together; others solve for a one week total. It provides an excellent class discussion and it is hoped that all groups will arrive at the same total.

Hint 1:

How far does he drive each day?

Hint 2:

How far does he drive each week?

The Long Stop

Davis is directly north of Castleton. Notus is directly east of Castleton. Homedale is 14 miles directly east of Davis. Vale is directly south of Notus, and Homedale is directly north of Vale. There are straight roads connecting all of these towns except Davis and Vale. It takes one hour to drive from Homedale to Notus without stopping provided you drive at 55 miles per hour. It took Dan two hours and 37 minutes to drive from Davis to Castleton at 55 miles per hour, because he had to stop on the way there.

How long would it have taken him if he had not stopped?

Comments:

The strategy here is to draw a diagram/map with the given information. Once this has been completed, the solution will be obvious.

Hint 1:

What do we need to know to answer the question?

Hint 2:

Draw a map showing the towns.

Baking Cupcakes
for the School

Mrs. Nakami's class decided to celebrate the common birthday of four of the members of their class by baking something for everyone in the sixth grade at their school. After some discussion, they settled on cupcakes. The class decided to make enough cupcakes so that every one of the 98 sixth graders in the school could have two. They bought cupcake papers, which come in packages of 30, to bake the cupcakes in.

How many packages of cupcake papers did they need?

Comments:

This problem involves dealing with remainders in a real life situation. Give assistance to those students who have difficulty with division, but who understand the problem and know what to do. Notice that the question is not how many cupcake papers but how many packages did they need?

Hint:

How many cupcakes do they need?

Variant:

(**) Suppose they bake cupcakes for five classes, each with 23 students?

Electric Bill

At right is a copy of our electric bill last May, and my Dad said we had to do something to cut down on our expenses. In fact, if we did not cut our electric bill in half, he wouldn't take any of us to Disneyland. So we all got busy and were able to cut the bill exactly in half. How many kilowatt hours of electricity did we use in June?

First 400 kwh @ 4¢	$16.00
Next 550 kwh @ 10¢	55.00
	71.00

(Note that the first 400 kilowatt-hours (kwh) are cheaper than the others; they only cost 4¢ each while the rest cost 10¢ each.)

Comments:

This problem is full of subproblems and it is necessary to read and understand all of the needed information. Once students know the dollar figure they're striving for, they should be able to focus on the needed information.

Hint:

How much was their bill in June?

Learning English

Gil, a young man from the island of Nuku Hiva, decides to study the English language. He studies for one year, gaining a vocabulary of 1,300 words. He vows to build his vocabulary to 8,000 words by the end of his second year of study. He begins to study very hard, and he learns 200 new words per month for the next seven months. He realizes that at his current pace he will fall short of his goal.

How many words must he start learning per month to reach his goal of 8,000 words by the end of his second year of study?

Comments:

The assumed information in this problem is that the student knows there are 12 months in a year. This will enable problem solvers to isolate the subproblem that concerns how many words per month for the last five months. Next assumption: same number each month?*

If this division problem gives your students difficulty, feel free to assist in any way you feel in appropriate.

*A student who loved patterns solved it by saying that Gil "got better every month so learned more words more easily".

Months in order	Words
1	460
2	760
3	1060
4	1360
5	1660

Hint 1:

How many words does Gil have left to learn?

Hint 2:

How much time is left?

Packing Books

When my older sister left for college this past September, she had a terribly difficult time packing everything. In addition to her clothes, her stereo and a box of dishes, she also had to take a pile of books. She put half of her books in one large box, one fourth of her books in another box, eight books on the floor of the back seat, and the remaining five books on the floor of the front seat.

How many books did she take back to college with her?

Comments:

There are two linked subproblems here: finding the number of books not in boxes, and finding the fraction of books not in boxes. The arithmetic may give some students difficulty. If so, let them describe to you how they would do the problem. Later, help them with the details.

Hint:

What fraction of her books were not in boxes?

Variant:

(**) The second box contains one-third of her books.

Digging the Ditch

My cousin got lucky last summer and got a job working on a ranch—or at least he thought he was lucky. Later he told me that his first job was to help dig a ditch 720 feet long to bring water to a small pond for the horses. He and two other men were digging the ditch, and the owner of the ranch said that if they kept digging steadily, they would be finished with the job in three days.

Unfortunately, after one day of digging, one of the other men hurt his back and had to do something else.

How much longer did it take to finish the job?

Comments:

Once students figure out the portion of ditch each man can dig in one day, the next subproblem becomes finding the length of ditch dug the first day or the number of man-days of work already done. Once these subproblems are solved, finding the time to dig the remainder of the ditch or the man-days required is straightforward.

Hint:

How much of the ditch can each man dig in one day or how many man-days of work will it take?

Variants:

(**)   The first man is hurt after one and one-half days of work.

(***) Give students the original question without specifying the length of the ditch. It is still possible to solve the problem.

Ice Cream

Susan's ninth birthday party with her five friends was a big success. Her mother had baked a cake in the shape of a teddy bear, and they had eaten three quarts of her favorite ice cream. Her older brother, Randy, told their mother that for his birthday he wanted seven friends and lots of ice cream "because my friends eat twice as much ice cream as Susan's."

If Randy is right, how much ice cream should his mother get.

Comments:

A group or two may get "stuck" if they disregard the *assumed* information: Susan ate ice cream, too.

Now it is SIMPLER as six other friends, including Susan, eat three quarts—and the same assumed information works for the Randy's friends: eight of his friends eat twice as much as eight of Susan's friends.

Hint 1:

How many people ate the three quarts?

Hint 2:

How many people will be eating ice cream?

Inspector Lee looked once again at the dead body of Horace Rimple and noted the time on the shattered watch: 1:10. The watch had undoubtedly been shattered by the bullet which killed old Mr. Rimple as he lay asleep in his bed.

"Tell me what happened today," he instructed the butler.

"As always, on Sunday, it was Mr. Rimple's wish to have the family gather for the afternoon. On days such as today, when he was feeling unwell, the members of the family were to go in and sit by his bed to talk, even if he looked asleep. He could still hear them even if he did not make the effort to open his eyes."

"Today, as customary, Robert spent the time from 1:00 to 1:03 with him; Susan the time from 1:03 to 1:06; James the time from 1:06 to 1:09; William the time from 1:09 to 1:12; Lawrence the time from 1:12 to 1:15; and Mary, the youngest, the time from 1:15 to 1:18. Mr. Rimple was very particular about those times. Each of them said he seemed to be asleep."

"Well," said Inspector Lee, "it seems very clear that William is the murderer."

"Unfortunately," said the butler, "it is not quite that simple. While Mr. Rimple loved that watch, it did not keep very good time. It lost exactly six minutes every 24 hours. I would set it correct each night at eleven before going to bed. Thus, the watch was always incorrect."

"Ah." said Inspector Lee. "Then the murderer must be . . . ."
Who?

Comments:

Once students have computed the correct time versus Mr. Rimple's watch's time, they need only to transfer that information to the times given in the problem. Some groups will need to work one hour at a time.

Hint 1:

How much time did his watch lose every four hours—every hour?

Hint 2:

What time is it really when Mr. Rimple's watch says 3 a.m.?

Pictures of a TV Star

My best friend, Mary, is crazy about Tom Selleck—and I do mean c-r-a-z-y. Her uncle works in the publicity department of the studio that produces "Magnum, P.I.," and every month or two he sends her a bunch of eight-by-ten inch photos which Mary promptly puts up on her picture wall. The last time I went over to her house, the whole wall, eight feet high and eleven feet wide, was covered with pictures in nice neat rows. Actually the whole wall wasn't quite covered. There was a one foot gap at the bottom and then a little space between separate photographs. But there were a lot of pictures of Tom Selleck. She told me how many, but I forgot. How can I figure it out?

Comments:

Your students will need to use a little common sense here to decide that the photos are usually ten inches high and eight inches wide. They might have to look at a photograph to decide. Then the question becomes how many ten-inch photographs can be put on a seven-foot wall. Similarly with width.

Hint 1:

How high is the picture area on the wall?

Hint 2:

How many rows of pictures can go up the wall?

Variant:

(***) There are 144 photographs on the wall. How much overlap on length and width?

I was reading a report on the dangers of smoking which talked about the increased possibilities of lung cancer, heart disease, strokes, and all sorts of other things. In fact, the author calculated that for each cigarette a person smokes, he loses ten *minutes* of life. My aunt smoked one pack of cigarettes a day for a year before she quit. Each pack had 20 cigarettes. If the author is correct, how many *hours* of her life did she lose?

Comments:

The topic of this problem is of obvious value itself. Solving it involves listing assumed information, 365 days in a regular year, and multiplying. The sub-problems and the order may differ from group to group, another interesting topic for class discussion.

Hint:

How many cigarettes did she smoke in a year?

The Dieters

● After a soccer game, I was discussing diets with Lisa, Mary Beth and Debbie. None of them really had to go on a diet but they all did. Lisa lost three more pounds than Debbie. Mary Beth lost four more pounds than I did and twice as much as Debbie. I thought that dropping from 107 to 99 pounds was fine; all I ever wanted was to get back below a hundred. Lisa only weighs 93 pounds now, but before her diet she used to weigh ___ . What?

Comments:

This is simply done as a series of subproblems: how much did the speaker lose? Mary Beth? Debbie? and finally Lisa. Each of the questions is easy, but some students get lost in deciding the order.

Hint:

How much did Mary Beth lose?

The Symgiaa Pyramid

In the center of the wall surrounding the largest building on Smygiaa, there is a pyramid made of bricks. It is 23 levels high. The lowest level, 23 bricks long, is a deep purple color. The second level, 22 bricks long is a lighter purple. As you count up, each level has one fewer brick than the one below, and the colors change as in the rainbow: purple to blue to green, etc.

How many bricks are in the pyramid altogether?

Comments:

This problem is considered of medium difficulty because it involves a pattern that needs to be worked out backwards. The pattern itself is simplistic. Some groups may wish to start by using fewer steps.

Hint:

How many bricks would be needed for a pyramid only two levels high? Three levels?

●   I was reading the results of the
200-yard backstroke in our local
paper, but one of the times was
missing. It said that the three
swimmers from Willamette High
had times of 2:12.43, 2:14.09 and
2:19.66, while from our school
Lovejoy had a time of 2:11.93,
and Horace a time of 2:17.59. I
knew that Fred Foster swam as
well, but they didn't mention his
time. All I did know from the other
information was that our team scored a total of nine points in the event
(where first place gets 6 points, second gets 4 points, third gets 3 points,
fourth gets 2 points and fifth gets one point).

●   In what place did Fred finish?

Comments:

Your group can solve this problem in a variety of ways—lists, charts, diagrams and other means. The strategies used should produce an interesting class discussion. This problem also provides ample opportunities for students to organize the needed information.

Hint:

How many points did Lovejoy score?

Mixing Punch for the Party

Susan arrived home from school to find a note from her mother taped to the refrigerator: "Attention: The first person home MUST help me get the punch ready for the party. Measure all of the punch mix in the bottle and add HALF that much orange juice to the bottle. DO IT NOW!"

So Susan did what the instructions said (there were two quarts of mix) and then went next door to play with Eileen.

Five minutes later, Paul came home, found the note and also followed the directions. Then he jumped on his bike to see a friend.

A few minutes after that Mrs. Pedretti came rushing in the house, saw the note still on the refrigerator and wailed, "Why do I have to do everything?". She also then proceeded to follow the directions on the note.

How much extra orange juice was in the punch?

Comments:

Students seem to like this problem, and it is not hard if they are willing to do it a step at a time. Students sometimes try to repeat the first measurement three times not realizing that the amount of liquid in the bottle changes each time. Help them to read the problem carefully so that they understand.

Hint:

How much liquid did Paul find in the bottle?

Variant:

(**) How much extra punch mix would be needed so the punch would be right?

The Kidnapping Caper

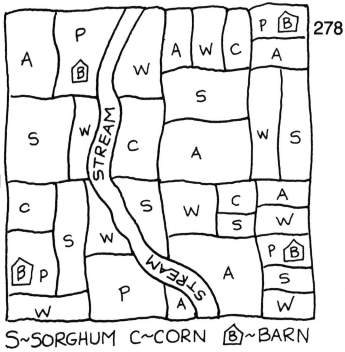

Inspector Lee said, "Tell me again, John what happened after you were kidnapped from your father's farm."

John said "Well, they blindfolded me and we drove for a long time. Then we got out and they took me through some fields to get to the hideout. It was a kind of curved path, but first we went through some alfalfa, then some wheat, some corn, some sorghum, then some more wheat and finally a pasture before they hid me in a barn. When they released me, they only took me about ten minutes away from the hiding place."

Inspector Lee said, "It's not too hard to guess where you were kept, then. Let me show you on this map." (Shown above.)

Where was John held prisoner?

S~SORGHUM C~CORN ⒷB~BARN
W~WHEAT P~PASTURE W~WHEAT

Comments:

The key to this is simply careful analysis beginning either with one of the barns and working backwards or else with one of the alfalfa fields and looking forwards.

Hint:

Find one barn John could have not been kept in.

The Compromise at The Conference

After a very long time for negotiation and compromise, the delegates from 23 nations decided that the principal conference would begin on the first Monday after the first Tuesday after the first Wednesday after the first Thursday after the first Friday after the first Saturday after the first Sunday after the first of January, 1983.

If January 1, 1983 falls on a Saturday, when did the conference begin?

Comments:

Working backwards, using fewer steps and making a list are strategies needed to solve this problem. Encourage your students to record their findings each step on the way.

Hint:

Make a chart showing the days and months, as needed, using the given information.

Motorcycle Racing

It is the seventh annual motorcycle cross-country race across the Nevada desert: 70 miles out to the flag and 70 miles back. Harry, on his new Harley-Davidson, averages 80 miles per hour going out but then has clutch trouble and can only manage 60 miles per hour coming back. Eric, on a Honda, can only go 70 miles per hour, but he keeps it up for the entire trip.

Who wins the race?

Comments:

This particular problem is often viewed as very easy and done incorrectly. It is a good lesson for checking your solution with the problem. The two boys *did not* tie!

Each time needs to be computed separately.

Hint:

How long did it take Harry to complete the race?

● Last year at Wembley Stadium in London, runners from all over the world gathered to break the world record for 100 miles. The leader from start to finish was the Ethiopian runner, Makla Flelera. He ran the first mile in exactly 7 minutes, the second mile in 7.01, the third mile in 7.02, and so on. After every 10 miles, he would sit down and do stretching exercises for 5 minutes and then resume his running.

How long did it take him to run all 100 miles?

Comments:

Once students recognize the pattern and the assumption that it holds true for the entire distance, the only remaining subproblem is figuring out stretching breaks and adding the totals together. See if they can MAKE THE PROBLEM SIMPLER by removing the first 7 minutes for each mile and adding the rest separately.

Hint:

How long did it take Makla to run the first 10 miles; the second 10 miles?

●   Our annual trip to see my uncle at Lake Tahoe was as dull as usual. Even rereading my favorite science-fiction novel wasn't helping. I was staring out the window when I saw a sign saying "Lone Pine—111 miles; Bishop—173 miles." That meant it was a long time until we had lunch at Bishop. I started reading again beginning with chapter nine and had just finished the tenth chapter when my mother said, "Look boys, it's exactly twice as far to Bishop as it is to Lone Pine."

How many miles had we driven while I read the two chapters in my book?

●

●

Comments:

A diagram will help your students decide how far apart the two towns are, which will lead to the distance of the car from Lone Pine when the narrator finished reading.

Hint:

How far apart are Lone Pine and Bishop?

Variant:

(**) Bishop is three times as far as Lone Pine.

The new sandbox at the nursery school is going to be 25 feet square and 2 feet deep, but only 18 inches of it will be filled with sand. One of the fathers in the group owns a pickup which can carry three tons of sand at a time.

If sand weighs 300 pounds per cubic foot, how many loads of sand will they have to bring in on the pickup?

Comments:

This problem requires an understanding of volume. It also assumes the knowledge that a ton is equivalent to 2,000 pounds. It serves as another example of a real life situation with no answer key.

Hint:

How many cubic feet of sand needs to be put in the sandbox?

Earthquake

A recent earthquake destroyed everything my rich uncle owned and made him sort of crazy. He took the six million dollars that he got from the insurance company and began spending it on having fun. The first day his fun only cost him $1; the second day $2; the third day $4; and so forth with the amount he spent doubling each day.

About how long will his six million dollars last?

Comments:

The problem provides a temptation for some students to simply multiply
and divide. You may wish to guide the students into using easier numbers
and solve first for $100. The new problem, also most easily solved by setting
up a pattern, should set most groups on the right track. Note that on the
last day he cannot double his spending from the previous day.

Hint:

How long will it take him to spend $100?

Going to the Museum

Our local Girl Scout troop was on its way to the local science museum. I always enjoyed these trips, especially since our leaders always took us on a route that went past two toy stores. That gave us a lot of different ways to get to the museum, even if we never went any extra blocks. There would have been even more routes if we hadn't agreed never to go by the corner of 14th and Market. That was where the Angel Lords hung out and they always hassled us. But we still had a lot of different routes. Can you tell how many?

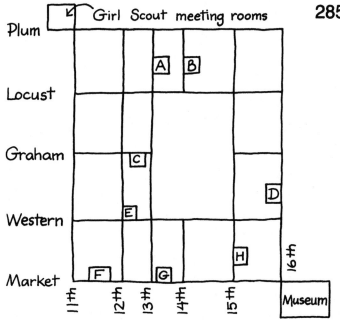

The toy stores are indicated by letters.

Comments:

Some careful analysis is the key to this problem. MAKE THE PROBLEM SIMPLER by selecting a first toy store to go by and counting those routes. A chart will surely be needed.

Hint:

Suppose you go by toy stores A and D. How many different routes are possible?

Carmine and her sister are selling raffle tickets for the swim team at the local supermarket. Carmine notices that about one person out of every fifteen who walk by their table stops for more information. Of the people who stop, approximately two thirds of them buy a $2 ticket. At the end of two hours, Carmine finds that they have sold $88 worth of tickets.

Approximately how many people walked by their table each hour?

Comments:

This is a fairly heavily disguised working backwards problem. It is probably easiest to compute that Carmine sells 22 tickets per hour, and observing that this represents two thirds of the people who stop. Now these people are one fifteenth of the total number who came by. Encourage your students to break up the problem into the simple subproblems.

Hint:

How many people stopped at Carmine's table each hour?

Variant:

Only one third of the people who stopped bought tickets.

The new Detroit radio station, WRCK, decided to stage a massive festival featuring the top 60 songs that week. The number one song that week would be played 60 times; the number two song would be played 59 times; the number three song would be played 58 times; and so on until the number 60 song would be played one time. Each song would take three minutes to play.

If the station also has 24 ten-minute news programs each day, how much time will be left for commercials during the week the festival is on?

Comments:

There are three parts to this problem. First, use the assumed information to get the total number of minutes per week. Second, subtract off the total number of minutes of news each week. Third, and most difficult, calculate the number of minutes that songs are played. Encourage your students to calculate the total number of songs played before multiplying by three.

Hint 1:

What is the hard part of this problem?

Hint 2:

Suppose only the top five songs were played: number five one time; number four two times; and so on until number 1 was played five times. Can you find a pattern?

Variant:

(*****) The number one song takes exactly 2 minutes; the number two song takes 2:01; the number three song takes 2:02; and so on; with the number sixty song taking 2:59.

Multiplying on a Watch

While waiting for her mother to finish shopping one afternoon, Jennifer saw that her new digital watch read 4:06:24, and realized 4 x 6 = 24. She began to wonder when this would happen again and was still staring at her watch when 4:07:28 appeared. Then she got a piece of paper and pencil out of her purse and began work in earnest to figure out how many times during an entire day that her watch would show a correct multiplication. Her mother dragged her away after she found 56 times, but she finished that night.

How many times did she find in all?

Comments:

Although this problem may appear difficult at first, patterning MAKES IT SIMPLER. Students will show and/or find these patterns in different ways. Allow time in class discussion for sharing.

Hints:

Let's start a list of all the times this works between four and five.

Variant:

(*) Jennifer divides the minutes by the hours to get the seconds.
(*) She multiplies seconds times minutes to get hours.

The keeper of the bird cages at the zoo discovered that two crested cockatoos would eat two pounds of bird seed every two weeks; that three Peruvian parrots would eat three pounds of bird seed every three weeks; and that four Mozambique macaws would eat four pounds of bird seed every four weeks.

How many pounds of bird seed will 12 crested cockatoos, 12 Peruvian parrots and 12 Mozambique macaws eat in 12 weeks?

Comments:

This problem consists of many subproblems, and the order and/or combinations do not matter. The various steps, hidden questions and hidden answers the individual teams come up with make interesting class discussion. Of note is that each type of bird eats a different amount of food per week. (See discussion on page 62.)

Hint:

How much seed does one Cockatoo eat in two weeks; in twelve weeks?

We were off on our first trip ever to Europe. Our plane left at noon, but the travel agent told us to be at the airport an hour ahead of time. So we all left the house at ten to drive the 40 miles to the airport.

However, for some reason there was very heavy traffic and my father could only average 20 miles an hour for the first 30 minutes. Both of my parents were getting more and more nervous before we got onto the freeway and the traffic cleared out. My mother asked me to figure out how fast we would have to go to get to the airport when the travel agent said.

What should my reply have been?

Comments:

Some prior experience with computing miles per hour is helpful in understanding this problem. Once the students have solved the subproblem of how many miles remain, the problem should be easily solved.

Hint:

How far have they traveled in the first 30 minutes?

Variant:

(***) The airport is 50 miles away.

Captain Spydyk was worried. A stray meteor had hit the Star Quest and severely damaged most of the air recycling system so that the carbon dioxide which the crew exhaled was slowly building up instead of being converted back into oxygen. Usually the equipment could convert 900 liters per hour of carbon dioxide back to oxygen, but now it was only possible to provide 150 liters of fresh oxygen per hour, while the crew needed almost 500 liters.

Captain Spydyk knew there were 6,500 liters of oxygen in reserve tanks and probably another 500 liters just in the interior of the ship. He also knew that if the total oxygen inside the ship was less than 200 liters, there would not be enough to breath.

According to his computer, the nearest planet with breathable air was 17 hours away. Could he get there safely? Prove your answer.

Comments:

A variety of subproblems need to be solved to answer the question thoroughly. Although the concept of rate is involved, the numbers used should be simple enough to provide successful completion.

Hint:

How much reserve oxygen will be needed each hour?

Variant:

Oxygen supply is below safety zone. What alternatives are available? (Science research involved.)

● It was hot driving across Arizona, but my father refused to turn on the air conditioning. "You know that if I turn on the air conditioning, we only get 16 miles per gallon. Right now, we're getting 18."

I thought that I was going to die from the heat when my mother said, "You know, George, turning on the air conditioner really isn't that expensive. And it's a whole lot cheaper than getting a divorce." My father took the hint and turned on the air conditioner.

● But I got interested and started figuring our how much it did cost per hour to run the air conditioner. We were going about 55 miles per hour out there in the desert, and gas cost $1.60 per gallon. How much did it cost us to keep cool?

Comments:

Your students should understand the concept of miles per gallon before attempting this problem. They will need to find out the cost per hour with the air conditioner off and on. The arithmetic gets messy so they may need to use easier numbers.

Hint:

How many gallons of gas will they use per hour if the air conditioner is on?

Computing Mileage

Yesterday when I stopped to fill up my diesel, I found that the pumps now registered in liters, although the price was posted as $1.12 per gallon. I wanted to compute the mileage my car was getting, but I could not remember the conversion factor from liters to gallons. All I did know is that I had driven 368 miles since the last fill-up and that the total bill was $14.28.

What mileage is my car getting?

Comments:

Given time, most students can break this problem down into the subproblem of finding the number of gallons of fuel purchased and then solving the problem. In one class, a fierce debate arose about whether or not you need to know the total capacity of the fuel tank. The teacher finally instructed them to make the computation using both a 20- and 30-gallon tank.

Hint:

How many gallons of diesel fuel were bought?

Variant:

(***) The price is $1.12 %/10.

My sister and I got lost in the mountains, but stumbled upon an abandoned miner's shack. Out behind the shack was an old well built in the shape of a square three feet on a side. There was some water in the well, but it was seven feet from the top, and we could only reach down about three feet. Luckily, my sister hit upon the idea of dropping some bricks into the well in order to raise the water level to a point where we could reach it. Each brick was about one fourth of a cubic foot in volume. How many bricks did we have to throw in to raise the water level up to where we could reach it?

Comments:

The solution to this problem requires an understanding of volume. If this understanding is present, then the subproblems about the number of cubic feet of water needed and their corresponding brick replacement can be solved!

Hint:

How much water would be needed to bring the level up to the girls' reach?

Variant:

(***) An ordinary building brick is 8" x 4" x 2 ½". How many of these would be needed?

● Last year my parents pulled me out of school in October to visit my grandmother in Vermont. It was a wonderful trip. The forests were a blaze of red and gold, and we even got to visit a neighbor who made maple syrup.

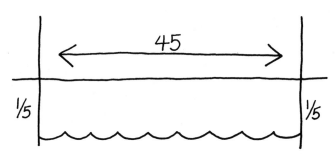

Just before we got to my grandmother's my father took a detour and we found one of the old covered bridges which still pop up from time to time. This one had one-fifth of its length over the west bank, one-fifth of its length over the east bank and the remaining 45 feet over the stream itself.

How long was the bridge?

●

●

Comments:

This is also a good problem for drawing a picture. A schematic diagram (as at right) will generally be more useful than a detailed picture. If you find two students with different types of drawings, try to get both of them on the board so that your students can see the advantages of each.

If the ratios are likely to give trouble, try the easier variant below.

45   ⅕   ⅕

Hint:

What fraction of the bridge is over the two banks together?

Variant:

(**) Here, two-fifths of the bridge is over each bank.

The Big Drawing

In order to raise money, the new Los Angeles professional soccer team, the Sea Lions, was having a lottery. Tickets were $2 each, they were able to sell 57,444 tickets. Out of this, they gave 202 prizes. First prize was worth $2,000; second prize was worth $1,500; third prize, $1,000; fourth prize, $995; fifth prize $990; sixth prize, $985; and so on with the value of each subsequent prize worth $5 less than the one before, until the 202nd prize was worth only $5.

How much profit did the club make from the raffle?

Comments:

Your students will need to recognize the pattern established here. Although it is possible to list all of the prizes and then total them, most students would not be willing to do so. Some guidance in using fewer steps may be needed, and/or you may need to deal with the obvious fact that the pattern is set up backwards.

Hint:

Create a new problem with 10 prizes, $50 for top prize and each successive prize $5 less than the last. What is the total prize value?

Horace Newport was sentenced to 12 years in jail for forgery. His forgeries were unfortunately detected rather quickly because he managed to misspell the word "dollars" (writing "dollers" instead [this actually happened]). To make the time pass more quickly, he has developed routines for morning, noon and night.

In the morning, after he gets out of bed, he dresses, makes his bed and cleans up his cell. At noon, he plays chess, reads the newspaper and goes for a walk around the exercise yard. At night, he gets undressed, brushes his teeth and hangs up his clothes. Each day he tries to do things in a somewhat different order than he did the day before.

If he does things in as many different ways as possible, how many days can pass before he has to start repeating himself?

Comments:

The solution to this problem is found simply by carefully reasoning. Students may or may not recognize it: diagrams, charts and/or lists of the many possibilities also provide the solution. Most students will break it down into smaller steps. Note that in the evening he must hang up his clothes after he gets undressed.

Hint:

How many days will be different if he is confined to his cell and is not allowed to do anything at noon?

Have you ever looked at your electric meter? There is a little dial that goes around once for each ten watt-hours of electricity used. When it goes around a hundred times, you have used one kilowatt hour of electricity.

One afternoon I was trying to figure out how much it cost to bake cookies, and I needed to know how many kilowatt-hours of electricity were used in ten minutes. With the oven turned off, the little dial on the electric meter took ten seconds to go around once. But when the oven was turned on, it only took the little dial six seconds to go around once. Then I was able to answer my question.

Show how I did it.

Comments:

The strategy here is one of isolating two subproblems and comparing their solutions. The students must also recognize the specialized meaning of *cost* in this problem. It refers to amount of electricity used, not dollars and cents.

Hint:

How many watt-hours of electricity would be used in ten minutes with the oven turned off?

Variant:

(***) Suppose that electricity costs 7¢ per kilowatt hour (a kilowatt hour is 1,000 watt hours). How much does it cost to use the oven for ten minutes?

Scale Model of the Solar System

As part of our astronomy unit, our class decided to lay out a scale model of the solar system in the local school gymnasium. After a lot of debate, we decided to include all nine planets. This meant that Pluto's orbit would be 180 feet in diameter, which was the width of the gym. We worked from the table at right giving the radius of the planets' orbits, and my job was to figure out what the radius of the Earth's orbit should be.

What answer did I get?

Planet	Distance from sun (in millions of miles)
Mercury	36
Venus	67
Earth	93
Mars	142
Jupiter	484
Saturn	884
Uranus	1789
Neptune	2809
Pluto	3685

Comments:

This problem assumes that students have some familiarity with ratio. Given that they understand the basic idea, most students find it helpful to use the strategy of make the numbers easier. Using estimation, students give Pluto a distance of 4,000 miles and Earth a distance of 100 miles. This strategy also helps them evaluate the original problem's solution.

Hint:

Suggest estimation as a technique and/or review ratio.

Variant:

Compute where all the planets should lie.

The Three Prisoners

●   There were three prisoners, one
of whom was blind. One day their
jailor offered to free them if they
could succeed in the following
game. The jailor produced three
white hats and two red hats and,
in the dark, placed a hat on each
prisoner. The prisoners were then
taken into the light where, except
for the blind man, they could see
one another. (None could see the
hat on his own head.) The game
was for any prisoner to state correctly what color hat he himself was wearing.
The jailor asked one of those who could see if he knew, and the man
answered no. Then the jailor asked the other man who could also see if he
knew, and his answer was no. The blind man at this point correctly stated the
color hat he was wearing, winning the game for all three.

What color hat was he wearing, and how did he know?

Comments:

This is an "oldie, but goodie". Students seem to have most success through experimentation and recording of all possibilities. It sometimes helps to have several students don paper hats and blindfolds in order to prove their solution.

Hint 1:

What are the possible combinations?

Hint 2:

Which ones should not even be considered?

This problem assumes that all three prisoners are fairly intelligent.

Suppose that the first man who is asked could see two red hats. Then he would know for sure that his hat was white. But he said that he did not know what color his own hat was. Hence, he could NOT see two red hats. So he saw either two white hats or a white and a red.

The second prisoner has figured out the above. If he sees two red hats, he also knows that his own hat is white. So, since he said "no", he does not see two red hats. So the second prisoner also must see two white hats or a red and a white hat. Furthermore, if he sees a red hat on the blind prisoner, then he knows that his own hat is white. (Recall what the first prisoner saw.) Since he did not know what color his own hat is, he could NOT see a red hat on the blind prisoner.

The blind prisoner understands all of this and knows that his own hat is white.

Problems to Illustrate Strategies★

Subproblems

111 – Same as 211
112 – Settling Up the Bills
113 – Raking the Yard
114 – The Candy Trove
115 – Snacks After School

211 – The Hrunkla Village
212 – Settling Up the Bills
213 – Washing Windows
214 – The Class Picnic
215 – New Kinds of Hot Dogs

311 – Same as 211
312 – Settling Up the Bills
313 – Folding Ads
314 – Eating Mints
315 – Washing Hair

Diagrams

121 – Same as 221
122 – Same as 222
123 – Dog, Goose, and Corn
124 – Arranging Coins
125 – Getting to Grandmother's
126 – Measuring with Sticks

221 – Commemorative Scroll
222 – The Circular Track
223 – Dog, Geese, and Corn
224 – Planting the Orchard
225 – The Three Squares
226 – Measuring the Paper

321 – Same as 221
322 – Capping the Bottles
323 – Crossing at White Falls
324 – The Temple's Cube
325 – Hrunkla Apartments
326 – Cooking Rice

Smaller Numbers

131 – Same as 231
132 – Princess's Wedding
133 – Halloween
134 – Same as 234
135 – Pocketful of Coins

231 – Spydyk and the Jewels
232 – The Goat Spell
233 – Stacking Cartons
234 – From the Black Lagoon
235 – A Purseful of Coins

331 – Same as 231
332 – Intergalactic Conference
333 – Driving Across the Desert
334 – Protozoa
335 – Mother's Day

Patterns

141 – Same as 241
142 – Same as 242
143 – African Spotted Mice
144 – Yellow Brick Road

241 – Handshake
242 – Space Station
243 – Teaching Reading
244 – Mailing Books

341 – Same as 241
342 – Same as 242
343 – Sumatran Lilies
344 – Spydyk's Bank Account

Working Backwards

151 – Same as 251
152 – Playing Investment
153 – Sale at Frozen Yogurt Shop

251 – Golden Apples
252 – Shopping Trip
253 – Lost in Desert

351 – Same as 251
352 – Coat on Sale
353 – The Bears

Multiple Solutions

161 – Mailing the Letter
162 – Crisis at Lunchtime
163 – Pancakes for Breakfast

261 – Freaking Out
262 – The Campout
263 – Mixing Feed in Nigeria

361 – After the Volleyball Game
362 – Carrying in the Bricks
363 – Lining the Field

★The hundreds' digit denotes the difficulty (100-level problems are easy, 200-level problems are intermediate, and 300-level problems are advanced), the tens' digit denotes the strategy emphasized, and the units' digit the suggested sequence.

Mind-Expanding Problems*

Easy

Medium

Difficult

---

*The numbers in parentheses after the title of the problem refer to strategies that may be useful in solving it:

0. Techniques for Understanding the Question
1. Subproblems
2. Diagram, Manipulatives
3. Simpler Numbers, Estimation
4. Patterns
5. Work Backwards